INNOVATION IN CHINA

China Today series

Richard P. Appelbaum, Cong Cao, Xueying Han, Rachel Parker and Denis Simon, *Innovation in China*

Greg Austin, *Cyber Policy in China*

Yanjie Bian, *Guanxi: How China Works*

Jeroen de Kloet and Anthony Y. H. Fung, *Youth Cultures in China*

Steven M. Goldstein, *China and Taiwan*

David S. G. Goodman, *Class in Contemporary China*

Stuart Harris, *China's Foreign Policy*

William R. Jankowiak and Robert L. Moore, *Family Life in China*

Elaine Jeffreys with Haiqing Yu, *Sex in China*

Michael Keane, *Creative Industries in China*

Joe C. B. Leung and Yuebin Xu, *China's Social Welfare*

Hongmei Li, *Advertising and Consumer Culture in China*

Orna Naftali, *Children in China*

Eva Pils, *Human Rights in China*

Pitman B. Potter, *China's Legal System*

Pun Ngai, *Migrant Labor in China*

Xuefei Ren, *Urban China*

Nancy E. Riley, *Population in China*

Judith Shapiro, *China's Environmental Challenges 2nd edition*

Alvin Y. So and Yin-wah Chu, *The Global Rise of China*

Teresa Wright, *Party and State in Post-Mao China*

Teresa Wright, *Popular Protest in China*

Jie Yang, *Mental Health in China*

You Ji, *China's Military Transformation*

LiAnne Yu, *Consumption in China*

Xiaowei Zang, *Ethnicity in China*

INNOVATION IN CHINA

Challenging the Global Science and Technology System

Richard P. Appelbaum
Cong Cao
Xueying Han
Rachel Parker
Denis Simon

polity

First published in 2018 by Polity Press

Polity Press
65 Bridge Street
Cambridge CB2 1UR, UK

Polity Press
101 Station Landing
Suite 300
Medford, MA 02155, USA

ISBN-13: 978-0-7456-8956-2
ISBN-13: 978-0-7456-8957-9(pb)

A catalogue record for this book is available from the British Library.

Library of Congress Cataloging-in-Publication Data

Names: Appelbaum, Richard P., author. | Cao, Cong, 1959- author. | Han, Xueying, author. | Parker, Rachel, 1971- author. | Simon, Denis, author.
Title: Challenging the global science and technology system / Richard P. Appelbaum, Cong Cao, Xueying Han, Rachel Parker, Denis Simon.
Description: Cambridge : Polity Press, 2018. | Series: China today | Includes bibliographical references and index.
Identifiers: LCCN 2018006109 (print) | LCCN 2018020432 (ebook) | ISBN 9780745689609 (Epub) | ISBN 9780745689562 (hardback) | ISBN 9780745689579 (pbk.)
Subjects: LCSH: Technology–China. | Science–China. | Research–China. | Technology and state–China.
Classification: LCC T173.5.C5 (ebook) | LCC T173.5.C5 A67 2018 (print) | DDC 338.951/06–dc23
LC record available at https://lccn.loc.gov/2018006109

Typeset in 11.5 on 15 Adobe Jenson
by Toppan Best-set Premedia Limited
Printed and bound in Great Britain by CPI Group (UK) Ltd, Croydon

For further information on Polity, visit our website:
politybooks.com

Contents

Table and Figures

Map

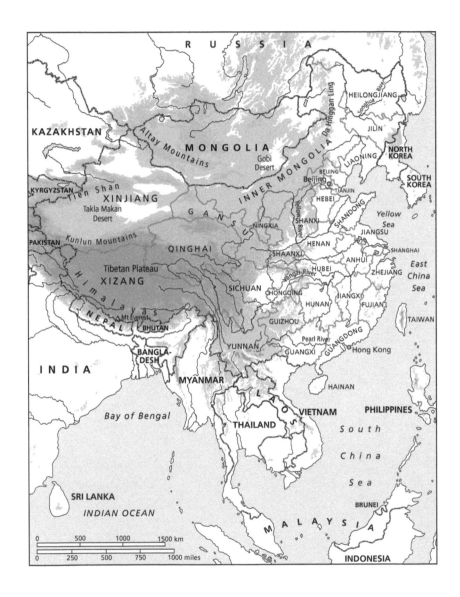

Chronology

1894–5	First Sino–Japanese War
1911	Fall of the Qing dynasty
1912	Republic of China established under Sun Yat-sen
1921	The Chinese Communist Party (CCP) established
1927	Split between Nationalists (KMT) and CCP; civil war begins
1934–5	CCP under Mao Zedong evades KMT in Long March
December 1937	Nanjing Massacre
1937–45	Second Sino–Japanese War
1945–9	Civil war between KMT and CCP resumes
October 1949	KMT retreats to Taiwan; Mao founds People's Republic of China (PRC)
November 1949	The Chinese Academy of Sciences (CAS) established
1950–3	Korean War
1953–7	First Five-Year Plan; PRC adopts Soviet-style economic planning
1954	First constitution of the PRC and first meeting of the National People's Congress
1956–7	Hundred Flowers Movement, a brief period of open political debate

1956	The Twelve Year (1956–67) Plan for the Development of Science and Technology launched, prioritizing scientific fields directly applicable to the development of nuclear weapons and their delivery
1957	Anti-Rightist Movement
1958–60	Great Leap Forward, an effort to transform China through rapid industrialization and collectivization
March 1959	Tibetan Uprising in Lhasa; Dalai Lama flees to India
1959–61	Three Hard Years, widespread famine with tens of millions of deaths
1960	Sino–Soviet split
1962	Sino–Indian War
October 1964	First PRC atomic bomb detonation
June 1967	First PRC hydrogen bomb detonation
April 1970	First PRC man-made satellite, "Dong Fang Hong 1," launched
1966–76	Great Proletarian Cultural Revolution; Mao reasserts power
February 1972	President Richard Nixon visits China; "Shanghai Communiqué" pledges to normalize US–China relations
September 1976	Death of Mao Zedong
October 1976	Ultra-Leftist Gang of Four arrested and sentenced
December 1978	Deng Xiaoping assumes power; launches Four Modernizations and economic reforms
1978	One-child family planning policy introduced

1979	US and China establish formal diplomatic ties; Deng Xiaoping visits Washington and US–China Agreement on Cooperation in Science and Technology is signed; PRC invades Vietnam
May 1980	Sino–Japan Science & Technology Cooperation Agreement signed
1982	Census reports PRC population at more than one billion
December 1984	Margaret Thatcher co-signs Sino–British Joint Declaration agreeing to return Hong Kong to China in 1997
1985	Reform of the Science and Technology System (S&T) starts
1986	The National High-Tech R&D Program (also known as the 863 Program) initiated; the National Natural Science Foundation of China (NSFC) established
1989	Tiananmen Square protests culminate in June 4 military crackdown
1992	Deng Xiaoping's Southern Tour re-energizes economic reforms
December 1992	Sino–Russia Science & Technology Cooperation Agreement signed
1993–2002	Jiang Zemin is president of PRC, continues economic growth agenda
March 1997	The National Basic Research Program (also known as the 973 Program) initiated
1997	Deng Xiaoping passes away; sovereignty over Hong Kong transferred to China

1998	The CAS launches the Knowledge Innovation Program (KIP); the Ministry of Education launches the World-Class University Program (also known as the 985 Program)
December 1998	EU–China Science & Technology Cooperation Agreement signed
November 2001	WTO accepts China as member
2002–12	Hu Jintao, General-Secretary CCP's Central Committee (and President of PRC from 2003 to 2013)
2002–3	SARS outbreak concentrated in PRC and Hong Kong
October 2003	Yang Liwei is put into orbit, marking the first success of PRC's Shenzhou Program, a manned spaceflight initiative
2006	PRC supplants US as largest CO_2 emitter; the Medium to Long-Term Plan for the Development of Science and Technology (2006–20) formulated, aiming to turn China into an innovation-oriented nation by 2020
August 2008	Summer Olympic Games in Beijing
2010	Shanghai World Exposition
2012	Xi Jinping appointed General-Secretary of the CCP's Central Committee (and President of PRC from 2013); new round of the reform of the S&T system starts
2013	"Made in China 2025" launched as an initiative to comprehensively upgrade Chinese industry
October 2015	Tu Youyou becomes first PRC Nobel Prize winner in science – Physiology or Medicine – for her discovery of artemisinin (also known as *qinghaosu*) used to treat malaria

December 2015	Consensus is reached in Paris within the United Nations Framework Convention on Climate Change (UNFCCC) dealing with greenhouse gas emissions mitigation, adaptation, and finance starting in the year 2020
August 2016	China's State Council issues national science and technology innovation plan during period of 13th Five-Year Plan (2016–20)
2017	Xi Jinping reappointed General-Secretary of the CCP's Central Committee (and President of PRC from 2018); the Program to Develop New-Generation Artificial Intelligence is unveiled
2018	National People's Congress removes two-term limit on China's Presidency, enabling Xi Jinping to remain President after his term would have expired in 2023

Abbreviations

AAAS	American Association for the Advancement of Science
ADU	Autonomous Driving Unit
AI	Artificial Intelligence
BERD	Business Expenditure on Research and Development
BRI	Belt and Road Initiative
BRIC	Brazil, Russia, India, China
CAS	Chinese Academy of Sciences
CAST	China Association for Science and Technology
CCP	Chinese Communist Party
CERC	Clean Energy Research Center
CNT	Carbon Nanotube
CSIA	China Semiconductor Industry Association
CSSIP	China–Singapore Suzhou Industrial Park
EPZ	Export Processing Zone
FDI	Foreign Direct Investment
FP7	Framework Programme 7
GDP	Gross Domestic Product
GERD	Gross Expenditure on Research and Development
H2020	Horizon 2020
IEEE	Institute of Electrical and Electronics Engineers
IoT	Internet of Things
IPR	Intellectual Property Rights

ITRI	Industrial Technology Research Institute (Taiwan)
JCM	Joint Commission Meeting
JICA	Japan International Cooperation Agency
JST	Japan Science and Technology Agency
LPWAN	Low-Power Wide-Area Network
MAU	Monthly Active User
MEMS	Microelectromechanical Systems
MLP	Medium to Long-Term Plan for the Development of Science and Technology
MNC	Multinational Company
MOE	Ministry of Education
MOFA	Ministry of Foreign Affairs
MOFCOM	Ministry of Commerce
MOFTEC	Ministry of Foreign Trade and Economic Cooperation
MOIIT	Ministry of Industry and Information Technology
MOST	Ministry of Science and Technology
MOU	Memorandum of Understanding
NAS	National Academy of Science (US)
NB-IoT	NarrowBand-Internet of Things
NDRC	National Development and Reform Commission
NIS	National Innovation System
NPC	National People's Congress
NPL	Nonperforming Loan
NSF	National Science Foundation
NSFC	National Natural Science Foundation of China
ODA	Office of Development Assistance
OSTP	Office of Science and Technology Policy
PCT	Patent Cooperation Treaty
PRC	People's Republic of China
PV	Photovoltaic
QUESS	Quantum Experiments at Space Scale

R&D	Research and Development
RFID	Radio-Frequency Identification
SAFEA	State Administration for Foreign Expert Affairs
S&ED	Strategic and Economic Dialogue
S&T	Science and Technology
SCI	Science Citation Index
SED	Strategic Economic Dialogue
SEI	Strategic Emerging Industries
SINANO	Suzhou Institute of Nanotech and Nanobionics
SIP	Suzhou Industrial Park
SIPO	State Intellectual Property Office
SME	Small and Medium-Sized Enterprise
SND	Suzhou New District
SOE	State-Owned Enterprise
STEM	Science, Technology, Engineering, and Mathematics
STI	Science, Technology, and Innovation
SUNY	State University of New York
USTR	US Trade Representative Office
VC	Venture Capital
VoC	Varieties of Capitalism
WIPO	World Intellectual Property Organization

Acknowledgments

This book is the result of a decade of research, primarily conducted under the auspices of UCSB's Center for Nanotechnology in Society (CNS), and supported by the US National Science Foundation (NSF) under Cooperative Agreements #SES 053114 and #SES 0938099. Research was also supported by the National Natural Science Foundation of China (NSFC) under grant 71774091. Special thanks also goes to Duke Kunshan University (DKU) for startup funding used to support research used in this project. Any opinions, findings, and conclusions or recommendations expressed in this material are those of the authors and do not necessarily reflect the views of the NSF, the NSFC, or DKU.

We are most appreciative of the NSF funding that made this research possible, and wish to single out Mihail (Mike) C. Roco, founding chair of the US National Science and Technology Council's subcommittee on Nanoscale Science, Engineering and Technology (NSET) and the Senior Advisor for the NSF's Science and Engineering at the National Science Foundation. CNS would not have existed were it not for Mike's vision of the transformative possibilities of nanotechnology; and without CNS, particularly the extraordinary leadership (and frequent hands-on advice) of CNS Principal Investigator and Director Dr. Barbara Herr Harthorn, our research would not have been possible. Others at CNS we wish to thank include CNS Assistant Director Bonnie Molitor, Center Administrator Shawn Barcelona, Travel and Purchasing Coordinator Valerie Kuan, and Education and Outreach Coordinator Brandon Fastman.

Our greatest debt of appreciation goes to the members and supporters of CNS's Interdisciplinary Research Groups on Globalization and Nanotechnology, which inevitably expanded to include a broader focus on China's high-tech turn. Our many research trips to China included visits and interviews at facilities in high-tech regions in Jiangsu Province (especially Suzhou Industrial Park), Shanghai, Beijing, Guangzhou, and Wuhan. Over the years, the members of our interdisciplinary research team who focused on China have included both graduate student and post-doc scientists and engineers. Special thanks go to two graduate students (now PhDs) who traveled with us to China in recent years, and played key roles in shaping our research: Matthew Gebbie in Materials, and Galen Stocking in Political Science. (Two of the co-authors of this book also began as graduate students – Rachel Parker in Sociology, and Xueying Han in Ecology, Evolution, and Marine Biology; after earning her PhD, Xueying continued as a postdoc.) We also owe a special thanks to Dr. Aashish Mehta (Global Studies), who contributed to our economic understanding, and to Luciano Kay, who joined us as a postdoc and subsequently as Research Faculty, providing expert data analysis. Among graduate students who in earlier years worked with us on the China project, we thank, on the science and engineering side: Peter Burks, Yiping Cao, Scott Ferguson, and Claron Ridge; on the social science side, Sarah Hartigan and James Walsh; and post-docs Stacey Frederick and Yasuyuki Motoyama. We were also fortunate to be able to draw on the expertise of UCSB science and engineering faculty, including Dr. Tim Cheng (Electrical and Computer Engineering and Dr. Bradley Chmelka (Chemical Engineering). We benefited from collaboration with Dr. Gary Gereffi (who joined us on an early trip to China) and Dr. Tim Lenoir (both Duke University), and Duke graduate student Patrick Herron. Phillip Shapira and Jan Youtie (Georgia Tech) worked with us to help develop our thoughts on the role of innovation in economic development and public policy. Finally, we wish to acknowledge several UCSB undergraduates who also contributed to our research efforts: Cece Choi, Andi Doktor, Emily Nightingale, and Joy Yang.

Finally, our thanks to Polity Press begin with Jonathan Skerrett, who recognized the importance of China's high-tech transition, proposed we write a book on the subject, and then stayed with us despite countless delays and lapsed deadlines. We also wish to acknowledge Amy Williams, who secured the manuscript reviews and initiated discussions about the book's description and cover; Karina Jákupsdóttir, who worked with us on the final preparation of the manuscript and figures, helping us to get the final version ready for production; our production editor Neil de Cort; Ian Tuttle, our meticulous and ever-patient copy-editor; and Adrienn Jelinek, who has the key role in our marketing efforts.

This book is dedicated to the pioneering American and Chinese scientists and engineers, whose cross-border research collaboration has provided the foundation for future S&T cooperation – cooperation that is badly needed to most effectively address the world's most pressing challenges.

Introduction: From the World's Factory to the World's Innovator?

China sees itself as an emerging world power. Shortly after his election as the General Secretary of the Central Committee of the Chinese Communist Party (CCP) in November 2012, during a visit to the National Museum off Tiananmen Square, Xi Jinping stood in front of an exhibit called "The Road to Rejuvenation," and reminded the assembled dignitaries and reporters:

> After the 170 or more years of constant struggle since the Opium Wars, the great revival of the Chinese nation enjoys glorious prospects … Now everyone is discussing the Chinese dream, and I believe that realizing the great revival of the Chinese nation is the greatest dream of the Chinese nation in modern times. (BBC News 2012)

Innovative breakthroughs in science and technology (S&T), resulting in globally competitive goods and services, are key to realizing Xi's Chinese dream. For decades China has been the world's factory, manufacturing and assembling goods ranging from cheap apparel to complex electronics. This role has served China well. Like the newly industrializing economies of East Asia a generation ago, China's meteoric economic growth has been fueled by export-oriented industrialization. The extraordinary amount of foreign direct investment China receives was initially driven by its plentiful supply of cheap labor. In recent years, however, as its economy has grown at historically unprecedented rates, its investments in science and technology have started to pay off.

Meanwhile, as labor costs have risen, foreign firms are increasingly drawn to China for reasons other than its manufacturing capabilities: access to what is becoming the world's largest consumer market, and the ability to partner with China's growing (and increasingly better educated) science and engineering talent pool.

China has grown to become the world's second-largest economy, its GDP surpassing Japan's in 2010. In the 33 years between 1978 and 2011, China's annual GDP growth averaged 10 percent, a 20-fold increase over the period (Haltmaier 2013). Even during the 2007–9 recession, China's economy only slowed to about 9 percent (World Bank 2013). Since that time, however, GDP growth has slowed further, as China makes the transition from public investment to consumer spending as a central driver of development. Still, China's GDP grew at 6.8 percent in 2017, more than three times the rate of the world's advanced economies (IMF 2017). When corrected for purchasing power parity, the IMF (2014) estimated that China's GDP surpassed that of the US in 2014.[1] China's rapid growth has resulted in a rising middle class, now estimated at several hundred million people.[2]

By October 2017 China's foreign exchange reserves had dropped to US$3.1 trillion, after reaching a high of US$4 trillion in the first quarter of 2014 (Trading Economics 2016; SCMP 2017); US$1.2 trillion (39 percent) were in the form of US Treasuries (US Department of Treasury 2017). China still retains a sizeable amount of foreign exchange reserves, which have been plowed back into the Chinese economy in the form of high-speed trains, highways, and other infrastructure; universities and science parks; and vast urban developments.

Public investment, which accounts for an estimated two-thirds of China's growth, has proven to be a successful strategy on the part of the Chinese Communist Party, whose legitimacy depends primarily on a rising tide that lifts a growing number of boats. But this approach has thus far produced limited returns in terms of innovation. China may be Walmart's leading trading partner, but the lion's share of the profits from that relationship is realized by Walmart. China may be the world's largest assembly of Apple and Samsung products, but only a small share of value added remains in China (see chapter 4 for further discussion).

As early as 2005, China's leaders set out to rectify this imbalance. China's role as the world's factory is undergoing a major transition: from "made in China" to "designed and created in China," from imitator to innovator. The 15-year Medium to Long-Term Plan for the Development of Science and Technology (MLP) set forth ambitious goals to transform China's S&T efforts from imitator to innovator (Ministry of Science and Technology 2006). It identified four basic science areas as "science megaprojects" (reproductive biology, protein science, quantum research, and nanotechnology, with stem cell and climate change added later), along with 16 "engineering megaprojects" (three being classified) and eight "frontier technologies" intended to convert scientific knowledge into commercially competitive leading-edge products.

The MLP has been backed up by China's 11th, 12th, and 13th Five-Year Plans, as well as a host of provincial and local efforts to develop world-class S&T capabilities. Significantly, the MLP emphasized the importance of "indigenous innovation" (*zizhu chuangxin*) capability to enable China to "leapfrog" its way into scientific leadership (the MLP is discussed in greater detail in chapters 2 and 3). Part of China's approach has been to move away from dependence on low-cost and resource-intensive production, which is seen as providing little in terms of technology transfer. Moreover, wages have been raised,[3] not only to placate workers, but also to promote the growth of a consuming class, having apparently learned from Henry Ford, who supposedly favored the "five dollar a day" wage so his workers could afford to buy the Model Ts they were producing. This, in turn, is resulting in some capital flight to lower-wage countries such as Vietnam, Bangladesh, Indonesia, and parts of Africa.

Since its adoption of the MLP over a decade ago, China's leaders have ramped up investment in "indigenous innovation" involving research, development, and commercialization of advanced technologies (Appelbaum et al. 2011a).[4] Although the MLP serves as an example of state-led S&T-oriented industrial policy, its investment in science and technology has yet to pay the predicted big dividends. As we shall demonstrate, although the trends are promising, China faces some significant barriers to achieving the world-class innovations it hopes to achieve. China's universities and science parks are impressive to look at, with laboratories and facilities that rival those of the US and Europe. Sparkling facilities, however, do not automatically translate into innovative breakthroughs.

In January 2013 China's State Council issued a notice advancing the government's plans for indigenous innovation.[5] The notice, an addendum to the 12th Five-Year Plan (2011–15) intended to support and implement the MLP, set forth State Council goals intended "to deepen the scientific and technological system to accelerate the nation's innovation system." It served as an official reminder that "the 12th Five-Year Plan states the urgent needs of our nation, at this crucial period, to build an innovation-oriented country, to thoroughly build a well and prosperous society, to accelerate the development of economic progress to enhance the capabilities of indigenous innovation to higher limits." The notice reaffirmed earlier calls for "supporting technological leapfrog development," to be achieved by "strengthening the building of basic conditions for technological innovation." These included an augmentation of research laboratory facilities, such as national key laboratories, improved instrumentation, and the construction of a network of field research stations. The notice called for "enhancing the ability to continuously innovate key industries," including advances in materials,[6] information technology, and energy.[7] There was a strong emphasis on green technologies, including the "implementation of a low-carbon technology innovation industrialization project, strengthening carbon capture, R&D and capabilities of utilization and sequestration technologies." Perhaps with growing traffic congestion and air pollution in mind, the notice also called for "accelerating the construction of intelligent digital traffic management, integrated transport, and green transportation." The notice called for a wide range of mechanisms to "strengthen the regional innovation development capacity [and] accelerate the construction of distinctive regional innovation systems." While the eastern region – already home to China's leading high-tech and manufacturing sectors – is to provide "an open resource-intensive sharing of its scientific and educational advantages," the notice also called for investment in central and western China as well. It specifically singled out "national high-tech industrial development zones," such as Beijing's Zhongguancun, Wuhan's East Lake, and Shanghai's Zhangjiang as the role models.

Finally, China's universities and vocational schools are seen as key to these efforts. While the notice called for such worthwhile goals as strengthening the educational system, enhancing people's abilities and talents, conducting world-class academic research, and promoting collaborative innovation, these are to be achieved largely through technological advances such as improved use of information technology and digital instruction, along with "gathering and training a group of top creative talents." There is no mention of more open approaches to instruction that might encourage creative, critical, "outside the box" thinking – an acknowledged shortcoming in China's quest to become a first-tier innovator. There is a call "to promote the popularization of science capacity building" through "science websites, virtual museums and virtual science museum building, the use of mobile phones, the Internet and mobile TV and other new media technologies and means of disseminating innovative scientific resources." The efforts of China's current leadership to boost innovation through these recent reforms will be taken up in chapter 3.

Is China becoming a high-tech world leader, leapfrogging into advanced technologies that will transform its economy into an "innovation powerhouse" – the title of a *Foreign Policy* article (Wertime 2014) typical of much recent hype? In this book we go beyond popular accounts, and dig more deeply into the evidence, drawing on extensive fieldwork in China that spans nearly three decades. We also examine the statistical evidence for China's advance, drawing on an original analysis of Chinese publications and patents in different fields, to determine the extent to which China's significant investments have paid off.

The first chapter, "China's Science and Technology Policy: A New Developmental State?", sets forth some of the principal issues that will be the focus of the book, centering on the evolving role of state policy as a key driver of both China's successes and failures. Our central question asks: Does China's state-led approach to science and technology constitute a new, more effective form of the developmental state? Or does the heavy-handed role of government impede truly creative thinking, discouraging innovation both in basic science and the resulting applied technologies? We begin by providing an overview of China's efforts to become a world power in science and technology, reviewing recent industry-sponsored studies that conclude that China's great leap forward in science and technology has been a resounding success. We then summarize the key points of China's current S&T plan for high-tech development. A final focus of this chapter is to evaluate the impacts of China's brain drain, and the effort of the government to convince the best and brightest overseas Chinese scientists and engineers to return through generous laboratory and financial incentives dubbed "talent programs" (for example, the Thousand Talents Program, the Young Thousand Talents Program). We evaluate the success of these efforts, arguing that the outcome has been, in fact, problematic (see also chapter 4). One unintended and undesirable consequence, for example, is a growing tension between those with and those without overseas experience.

In chapter 2, "Science and Technology in China: A Historical Overview," we survey the development of science and technology in China, providing a historical context to better understand China's current efforts. We begin with a description of the process through which modern science was first introduced and institutionalized in China, before turning to a more extensive discussion of the organization of scientific research in the People's Republic of China (PRC), beginning with the eras of Mao Zedong and Deng Xiaoping, up through the MLP and emphasis on indigenous innovation. We examine the intrusion of politics into science and research – China's top-down approach in formulating and implementing S&T and innovation policy, a historical legacy that has had its enduring impacts. We conclude the chapter with a discussion of the challenges China faces in becoming an innovation-oriented nation and a world leader in science and technology – a topic to which we return throughout the book.

In chapter 3, "China's Science and Technology Enterprise: Can Government-Led Efforts Successfully Spur Innovation?", we return to our key concern by evaluating the role of the Chinese government as a driver of science and technology development in such key fields as life sciences, nanotechnology, advanced manufacturing, aerospace, clean energy, and supercomputing. The Chinese Communist Party continues to control the scientific enterprise through a Leading Group on Science, Technology, and Education at the State Council and other mechanisms. The central government has initiated various top-down national programs, including the previously mentioned talent-attracting programs, and has approved the setting up of 168 national high-tech parks. The central government still controls a significant share of the R&D funding, which is often distributed without due attention to merit or transparency; its use has all too often been ineffective, inefficient, wasted and abused. We take a critical look at state-led efforts, including the growing emphasis on science and technology parks as innovation hubs.

In chapter 4, "China's International S&T Relations: From Self-Reliance to Active Global Engagement," we document and analyze China's rapidly evolving international S&T relations – its cooperation, collaboration, and occasionally adversarial relations with its neighbors in Japan and Korea, as well as with the West. Since the announcement of the S&T modernization program in the late 1970s, China's leadership has believed that international engagement would be an important vehicle to help the Chinese S&T system catch up with the West. As a result, Chinese governments at the national, provincial, and local levels have signed a broad range of bilateral and multilateral S&T cooperation agreements with the world's leading nations. How effectively are these relations playing out? How have they shifted as a result of the strengthening of China's own S&T capabilities? In addition to looking at a select array of specific governmental relationships, we also examine the role of technology imports in supporting China's S&T advance, examining the PRC leadership's growing concern about the country's continued high degree of dependence on foreign know-how to drive its economic development. This chapter also analyzes the growing presence of foreign R&D centers in China, which now number over 1,800, as well as the emergence of Chinese global firms and their efforts to extend domestic R&D centers in addition to "listening posts" abroad.

In chapter 5, "How Effective Is China's State-Led Approach to High-Tech Development?", we address the question of whether high-tech parks can be a key to promoting innovative breakthroughs in China. We first briefly discuss China's current model of economic growth, challenges, and high-tech turn, after which we examine its often contradictory blend of heavy-handed state-driven development and untrammeled free enterprise. After an examination of nanotechnology as one area in which China has sought to achieve indigenous innovation, we then focus on Suzhou Industrial Park (self-characterized as and competing for being "China's Silicon Valley"[8]), which has an annex devoted to nanotechnology. Dubbed "Nanopolis" (a play on Singapore's successful Biopolis), this annex is home to some of China's nanotechnology startups, often involving international partnerships through former Chinese expats (now returnees) with Silicon Valley connections. Firms are provided support for business plan development, legal and incubation services, tax holidays, and other perks. In this chapter we illustrate how China's approach is seeking to create a new regional economic advantage for the country, one that is based on high-tech innovation through nanotech-related firms and supporting institutions. We conclude with some reflections on China's high-tech future, and the successes (and failures) of China's policies intended to foster innovation-driven economic development.

In our final chapter, "Xi Jinping's Chinese Dream: Some Challenges," we ask the question: Is there a Chinese model of innovation? China has broad ambitions and has invested considerable public resources in a well-orchestrated effort to achieve its goals. Can China's industrial and/or innovation policy pay off? Is a Nobel Prize around the corner? We conclude with some speculations on what China's high-tech ambitions, if successful, might mean for the Chinese economy as well as for its relations with the rest of the world. What are the challenges that China will face, if it succeeds in moving away from an economy based on export-oriented industrialization and public investment, to one that emphasizes technological prowess and private consumption? How will this impact China's growing industrial workforce and its talent pool? Finally, to the extent that China succeeds in its efforts at indigenous innovation, relying on its own vast and growing middle class as consumers for Chinese brands, how will this affect its economic and political relations with the United States and other countries?

<table>
<tr><td>1</td><td>China's Science and
Technology Policy
A New Developmental
State?</td></tr>
</table>

China's Science and Technology Policy: A New Developmental State?

1

Among its various goals, China aims to become an "innovation-oriented" society by the year 2020 and a world science and technology superpower by 2050; "indigenous innovation" (*zizhu chuangxin*) is seen as the source of China's future development. Through its fifteen-year Medium to Long-Term Plan for the Development of Science and Technology (MLP) and a series of five-year plans, the Chinese government has sought to move away from the export-oriented manufacturing that provided the basis for double-digit economic growth for nearly three decades, instead focusing on STEM (science, technology, engineering, and mathematics) education, home-grown innovation, and fostering commercialization through national, state and local government investments that range from support for promising projects to entire high-tech parks.

CHINA'S SUCCESSES: THE VIEW FROM OUTSIDE

Based on a variety of reports that seek to measure global competitiveness and innovation, China is a rising star. One survey of the top executives of foreign firms operating in China found that two-thirds of the executives believed that Chinese competitors were at least as innovative – and in some cases more innovative – than their own firms in China. Chinese companies were seen as outpacing their competitors in such high value-added activities as "advanced and applied research, as well as emerging technologies and trend analyses" (Veldhoen et al.

2014: 6).[1] The survey also reported, however, that innovative success was due in large part to the acquisition of foreign technologies. Other studies differ, however: the Boston Consulting Group's 2016 list of the world's 50 most innovative companies includes only two Chinese firms, Xiaomi (ranked 35th) and Huawei (ranked 46th) (BCG 2017a, 2017b).[2]

The World Economic Forum's *Global Competitiveness Report 2016–2017* ranks China 28th, unchanged from the previous year. Its measure of competitiveness – defined as "the set of institutions, policies, and factors that determine the level of productivity of an economy, which in turn sets the level of prosperity that the country can achieve" (WEF 2016: 4) – is based on a set of weighted indices and subindices, with data derived from the IMF, World Bank, and UN agencies such as the International Telecommunication Union, UNESCO, and WHO;[3] the indices also draw on an Executive Opinion Survey of nearly 150,000 business executives in 141 economies, which resulted in nearly 14,000 usable responses (WEF 2016: 77). Of the various measures used, two are seen as "key for innovation-driven economies:" business sophistication and innovation, with China ranking 34th on the former and 30th on the latter (WEF 2016: 148–9).[4] The report concludes that China has made some progress

in some of the more sophisticated areas of competitiveness that contribute to shaping the country's innovation ecosystem. These include higher education (54th, up 14), innovation (30th, up one), and business sophistication (34th, up four). This bodes well for the future while China transitions to a new normal, where growth will need to be increasingly driven by innovation. Yet China still lags behind in technological readiness (74th, unchanged) despite a significant improvement in all components of this category since last year… [but] gains posted in these categories are partially offset by a worsening fiscal situation [and] inefficiencies and instability characterize the financial sector. (WEF 2016: 146)

The *Global Innovation Index 2016*, the result of a collaboration between Cornell University's College of Business, INSEAD, and the World Intellectual Property Organization (WIPO), gives China significantly higher marks than the World Economic Forum study: it places China 25th among the 128 countries that are ranked in terms of 82 indicators of overall innovation; in terms of innovation quality – based on such metrics as university rankings, international patent filings, and publication citations, China placed 17th, making it the first middle-income economy to move into the top half of 49 high-income economies (WIPO 2016a).[5] China's standing reflects the fact that "the country has a particularly high number of R&D-intensive firms among the top global corporate R&D spenders. China's innovation rankings this year also reflect high scores in both the Business sophistication and Knowledge and technology outputs pillars, in which it scores above the average of the overall ranked 11–25 group to which it now belongs" (Dutta et al. 2016: 10).[6]

KPMG's September–November 2016 survey of 841 high-tech executives across 15 countries reported that roughly a quarter of all respondents believed that within four years China – rather than the United States – would hold the greatest promise for creating disruptive technology breakthroughs that would have a global impact (roughly the same number – 26 percent – favored the United States, a decline of 3 percent from the previous year). Shanghai was predicted to become the world's leading high-tech innovation hub (KMPG 2017; Perez 2017).[7] Interestingly, there was considerable divergence of opinion, based on whether the respondent was from the US or China: 49 percent of US high-tech executives favored the United States, while 59 percent of the Chinese respondents favored China. Such apparent nationalistic bias was also found when the results were broken down by region (KMPG 2017: 3).[8] China was seen as outperforming the United States in mobile commerce, innovative financial information technologies that utilize big data, risk modeling, cloud computing, and blockchain modeling, leading the KMPG report to conclude:

> China is transforming from an investment-intensive, export-led model of growth to one driven by consumption and innovation. Through the use of disruptive technologies such as cloud computing, the Internet of Things, smart industrial robotics, Data & Analytics, and enhanced automation, Chinese companies are capturing new business opportunities ... China's CEOs identified new product development, increasing data analysis capabilities, the Internet of Things, machine-to-machine technology, and the industrial Internet as top focus areas for further investment in the next three years. (KMPG 2017: 16)

While China's rankings are strong among emerging economies, as we shall argue throughout this book, its goal of becoming a world leader in innovation remains challenging.

INNOVATION: THE STATE'S GOALS

Alongside sustained and unprecedented economic growth, China has focused state-led efforts on ensuring that science, technology, and innovation (STI) propel the country as a key growth engine (OECD 2017a). During the past two decades, China's research and development (R&D) achievements have been steadily increasing. R&D intensity, measured as gross expenditure on R&D (GERD) as a percentage of GDP has gone up from just under 0.9 percent in 2000 to 2.1 percent in 2016 (OECD MSTI 2017). China's state-led focus on policy reforms and plans, including the *13th Five-Year Plan on Science, Technology and Innovation*, establishes a series of policies and targets for the country's S&T ecosystem to achieve during the 2016–20 period (OECD 2017a). China has set ambitious targets to building technological capacity in a range of sectors including ICT, robotics, new materials, and clean energy, amongst others (OECD 2017a). During this period, the overall growth rate of R&D in China between 2001 and 2011 reached more than 20 percent annually (NSF 2014). Business expenditure in research and development (BERD) reached 1.4 percent in 2012 in China compared to 1.3 percent across Europe. China has continued to increase its R&D spending and is estimated to overtake both the EU and the US to become the world's top R&D spender by 2019 (OECD 2014).

At the 2014 World Economic Forum summer meeting in Tianjin, Premier Li Keqiang formally announced China's "mass entrepreneurship and mass innovation" campaign, involving central government financial support for startups. In that year China had registered approximately 5.1 million new enterprises, at an average of 14,000 per day. By 2017, the number of newly registered enterprises had increased to approximately 16,000 per day, totaling 5.8 million new enterprises for the year (Xinhua 2017a, 2017b). In the absence of independent performance data, however, it is difficult to assess the success of Li's campaign – to know how many of these firms are innovative, or even how many have succeeded. Of high-tech startups among recent university graduates, for example, Xinhua (China's official news agency) reported a success rate of only 2.4 percent (He 2017). Yet although China is reportedly the world's second largest market for venture capital, it has created relatively few internationally known startups worth more than a billion US dollars – the so-called "unicorns" (Yiu 2016) that are often cited as indicators of successful innovation (we briefly review the success of several "unicorns" in chapter 5).

China's *13th Five-Year National Plan for Science, Technology, and Innovation* (2016–20), issued in August 2016, identified what it regarded as

> significant innovative achievements in manned spaceflight, lunar exploration, manned deep diving, deep drilling, supercomputing, quantum anomalous Hall effect, quantum communication, neutrino oscillation and induced pluripotent stem cell, etc. In 2015, the R&D expenditures

in China reached 1.422 trillion yuan [US$228.3 billion]; the number
of international scientific papers ranked the second place in the world
and the number of citations of these papers ranked 4th in the world;
the total value of contract deals in domestic technical markets reached
983.5 billion yuan nationwide [US$157.9 billion]; and the national
capacity in innovation ranked 18th in the world.... Major breakthroughs
have been made in high-speed railways, hydro power equipment, UHV
AC power transmission and transformation, hybrid rice, the fourth
generation mobile communication (4G), earth observation satellite,
Beidou navigation system, electric vehicles and other major equipment
and strategic products.... Big leaps are being made in such fields as
information technology and networks, artificial intelligence, biotechnology, clean energy, new materials and advanced manufacturing. Disruptive technologies are emerging. (China STI 2016: 2–3)

Notwithstanding this bullish appraisal, the Plan also acknowledges
that

China still has some weak links and underlying problems in STI, such
as weak scientific and technological bases, big gap between China and
the developed countries in the innovation capacity, especially the original
innovation capacity, the basically unchanged situation that core technologies in key areas are still controlled by other countries, many
industries are remaining in the low and medium ends of the global
value chain, and the contribution rate of science and technology to the
economic growth is not very high. (China STI 2016: 4–5)

The Plan proposes to remedy these deficiencies, emphasizing the
original MLP's call for indigenous innovation in the hope of elevating
China to become "one of top 15 nations in the world with regard to
overall innovation capacity" (China STI 2016: 8). Characteristically,
the Plan sets specific and ambitious metrics for success (China STI
2016: 10) (see Table 1.1).

Table 1.1 Major Indicators of STI in the 13th Five-Year Plan Period*

Indicators		2015	2020
1	World rank in overall national innovation capacity	18	15
2	Contribution of S&T progress to economic growth (%)	55.3	60
3	Ratio of gross R&D expenditure to GDP (%)	2.1	2.5
4	Number of R&D personnel in every 10,000 employed people	48.5	60
5	Revenues of high- and new-technology enterprises (trillion yuan)	22.2	34
6	Ratio of the added value of knowledge-intensive services to GDP (%)	15.6	20
7	Ratio of R&D expenditures to main business revenues in industrial enterprises above the designated size (%)	0.9	1.1
8	World rank in the number of citations to China's publications in international journals	4	2
9	Number of patent applications submitted through the Patent Cooperation Treaty (PCT) (in 10,000s)	3.05	Double
10	Number of patents for inventions owned for every 10,000 people	6.3	12
11	Value of contract deals in domestic technical markets (billion yuan)	983.5	2000
12	Percentage of citizens with scientific literacy (%)	6.2	10

*Corresponding figures, in dollars: row 5 = US$3.5 trillion, US$5.5 trillion; row 11 = US$157.9 billion, US$321.1 billion. All yuan-dollar conversions are in current US dollars, using the China Statistical Yearbook (2017) yearly conversion tables
Source: Compiled with data from China STI (2016)

Mechanisms for achieving these goals include building city, regional, and provincial innovation centers; creating national high-tech zones and regional clusters; strengthening the rule of law and IP protections; and exploring "a new type of centralized system for STI in a socialist market economy" (China STI 2016: 16). Major national S&T high-tech projects are to include electronics, cloud computing, nanoscale lithography chip etching, 5G mobile communications, and high-end numerical control machines.[9] Projects earmarked for 2030 include quantum communication and quantum computing, brain science and artificial intelligence (AI) research, green development of coal, smart grid, intelligent (3D, robotic) manufacturing, development and application of key new materials. The Plan specifies, in great detail, how these objectives will play out in terms of specific technological applications, with nanotechnology reaffirmed as a key driver of innovative breakthroughs (as it was in the MLP):

Develop new functional nano-materials, nano optoelectronic devices and integrated systems, nano biomedical materials, nano-drugs, nano energy materials and devices, nano environmental materials, nano security and detection technology and so on, make breakthrough in key technologies and standards on the preparation of nano materials and processing of devices, and strengthen demonstration and application. (China STI 2016: 32–3)

The 13th Five-Year STI Plan emphasizes "talent development as a 'priority' strategy" (China STI 2016: 7), calling for strengthening "the training and use of innovative scientific and technological talents from minority groups, and value and increase the proportion of female scientific and technological talents" (China STI 2016: 69). This is to be achieved both by educational reforms and by

expand[ing] the introduction of overseas high-caliber talents. With a focus on the major national demands, introduce high-level innovative

Box 1. Quantum Communication

One area where China is making a strong effort to become a world leader, singled out in the Plan, is quantum communication. Quantum research was listed as one of four science megaprojects in China's MLP; in 2016, quantum communications and quantum computing were listed as a key national strategic industry; and quantum communications was designated as one of six major S&T development projects for 2030 in China's 13th Five-Year Plan (National People's Congress 2016). China has made significant strides in this field and made worldwide news on August 16, 2016 when it launched the world's first quantum satellite, named Quantum Experiments at Space Scale (QUESS) (Aron 2016). QUESS will allow researchers to test theories regarding the quantum property of entanglement, which states that two particles can be "entangled" to one another no matter the distance that separates the particles. Theoretically, quantum cryptography provides the only truly secure method for transmitting encrypted messages. A property of entanglement states that when one of the entangled particles is acted upon by an external force, such as a hacker, both particles are affected, which not only guarantees that the message itself is no longer valid as the particles carrying the information have been altered, but it also alerts the original messengers that there was an attempt to intercept the message. Over the next two years, QUESS will attempt to successfully send secure messages between Beijing and Urumqi. The satellite uses an ultraviolet laser to generate entangled pairs of photons, which will then be directed to the two base stations (Aron 2016; BBC News 2016).

The ability to send secure messages using quantum properties has become increasingly relevant and necessary as more headway on making stable, larger qubits (the quantum analogue of a classical bit) is made. With the promise of quantum computers being available in the near future (Hackett 2017; Johnston 2017), classical encryption schemes based on mathematical algorithms become increasingly vulnerable and many companies are already looking for post-quantum cryptography methods to guarantee the secure transfer of data (Cloud Security Alliance 2017). In addition to QUESS, China is also building the world's longest ground-based quantum encryption network to link the 1,200 miles between Beijing and Shanghai (Chinese Academy of Sciences 2016). Unsurprisingly, China is also the global leader in scientific publications relating to quantum communications, followed by the United States, the United Kingdom, Germany, and Japan.[1]

[1] Based on a SCOPUS search conducted by the authors. Search query: TITLE-ABS-KEY ("quantum communication" OR "quantum communications" OR "quantum key distribution" OR "QKD" OR "quantum cryptography"). Publications were limited to articles published between 1965 and 2017. Search was performed on March 21, 2017.

talents from the rest of the world, such as chief scientists and special talents urgently needed by China, establish dedicated channels, implement special policies, and realize accurate introduction. Improve working and living environment and related services for foreign experts. (China STI 2016: 70)

The Plan is indicative, not concrete: it sets multiple priorities, but does not specify the details for implementation. Nor is the issue of funding addressed: the Plan does not say how much will be spent to achieve its objectives, nor where funding will come from (central, provincial, local authorities or enterprises that presumably will contribute the majority of the funding). The details for implementation will be worked out, although in an atmosphere of slowed economic growth that will likely affect the Plan's ability to achieve its lofty goals.

SCIENTIFIC PUBLICATIONS: QUANTITY OVER QUALITY

The combination of China's financial support for R&D, sustained attention to science policy (including continuous reform efforts), and the sheer size of the S&T talent pool have led to explosive growth in scientific output (figure 1.1). During the 1990s, Chinese publications lagged behind the United Kingdom, Germany, Japan, and the United States. As China renewed its commitment to R&D, there was a noticeable upsurge in the country's scientific publications beginning in the early 2000s (Xie et al. 2014) with key gains in fields such as Engineering, Physics & Astronomy, Materials, and Chemistry, in all of which China out-produced Europe in terms of the overall share of academic publications. The growth rate in Chinese-authored scientific publications, in some fields, has reflected sustained double-digit

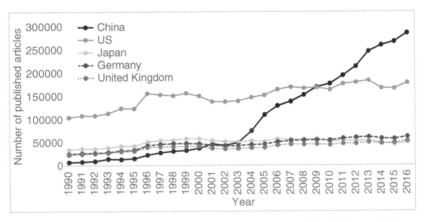

Figure 1.1 Total Number of Scientific Articles Published from 1990 to 2016 for Selected Countries

Source: Compiled with data from Scopus, focusing only on scientific articles published in the physical and life sciences

increases. In 2009, China overtook the US as the world's leading scientific publisher,[10] at least in terms of the number of SCOPUS-indexed publications.[11]

International collaboration and co-authorship are known to yield more high impact and more highly cited publications and will be crucial for China's future success. Multiple studies have documented China's increased participation in international collaboration (Fan et al. 2014; Sergi et al. 2014; Ministry of Science and Technology 2014; Bornmann et al. 2015) as a key indicator of increased scientific impact overall. Collaboration for "mutual advantage" through research partnerships is on the rise in China, which now collaborates with the US and the UK more than with any other countries, globally (Bound et al. 2013). The rise in collaborative research is a result of China's new policies and funding schemes aimed at encouraging the exchange of knowledge domestically, but there is also evidence that OECD countries (amongst others) are increasingly interested in international collaborations with China.

Several national funding agencies have set up "overseas offices" in China, including the US National Science Foundation, the Research Councils UK, and the German Research Foundation (DFG). Such offices play a central role in facilitating international connections that are both strategic at the national level but also practical at the level of individual researchers, making introductions and helping foreign researchers to gain entry into China (Sergi et al. 2014). As overall publication output continues to grow, the enabling benefits of international collaboration through overseas offices, such as that of the US National Science Foundation in Beijing, have been seen as playing a central role in facilitating not only an increase in publications, but in raising quality and therefore impact of co-authored publications between Chinese researchers and foreign counterparts. The US, however, has recently decided to close its NSF office in Beijing (as well as other overseas offices), in an effort to become more internationally "strategic and focused" (Normile and Stone 2018) – a decision which will likely impede future international S&T cooperation.

CHINA'S SCIENTIFIC TALENT POOL: COMING OR GOING?

China's educational system still lags behind Europe and the United States (Ministry of Science and Technology 2014). Nonetheless, with its increasing research funding and its better career opportunities for researchers, China has sought not only to increase the size of the country's scientific talent pool, but the quality of its output as well. The Cultural Revolution (1966–76) was a significant setback for China's educational system: national college entrance examinations were halted, no new students were admitted to institutions of higher education, and a generation of scientists was effectively lost (Simon and Cao 2009a). Since 1978, however, China has made significant gains. Many of the S&T policies that were adopted to increase China's international competitiveness have been focused on institutions of higher education (Li 2010). The rapid expansion of China's higher education system, which began with the 1978 economic reforms and especially amid the Asian financial crisis in the late 1990s, has seen undergraduate enrollment in regular higher education institutions increase 30-fold between 1978 and 2015 (from 856,000 students to 26 million) and graduate student enrollment increase by a factor of 160 (from 12,000 to 1.9 million students) (Ministry of Education 2016; see also Li 2010). Education has always been highly valued in China and is still viewed as a primary way to increase one's social standing (Appelbaum et al. 2016). The effort, particularly in regard to increasing the number of PhD graduates, has paid off. In 2005, China awarded 27,677 PhD degrees, 15,978 (58 percent) of which were in an S&T-related field (China Statistical Yearbook 2006) (figure 1.2). By 2015, the number of PhD graduates had almost doubled. In 2015, China awarded 53,778 PhD degrees,

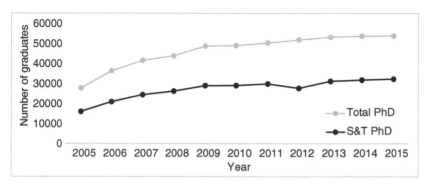

Figure 1.2 Total Number of PhD Graduates across All Disciplines and for S&T Fields in China from 2005 to 2015
Source: Compiled with data retrieved from China Statistical Yearbooks 2006–16.
Note: S&T fields consist of those categorized as science, engineering, and agriculture academic fields in the Statistical Yearbooks. See References for full citations.

32,256 (60 percent) of which were in an S&T-related field (China Statistical Yearbook 2016). The total number of PhDs granted in China increased by more than 75 percent from 2005 to 2009 but has plateaued over the last several years.

In addition to domestically trained PhD graduates, China also has a large pool of external talent – individuals who pursued graduate degrees abroad to potentially tap into through collaborative research. For example, in the United States alone, Chinese students accounted for 16 percent of all US S&T doctoral recipients in the 2015–16 academic year (NSF and NCSES 2016). Overall, the number of Chinese students studying abroad has increased over time, reaching 544,500 in 2016 (see chapter 4 for further discussion). The United States continues to be the primary destination for Chinese students (Neubauer and Zhang 2015; Colson 2016; McPhillips 2016), and China has remained the top sending country for the US since the 2009–10 academic year (IIE 2010, 2011, 2012, 2013, 2014, 2015, 2016) (figure 1.3). The number of Chinese students studying at US institutions of

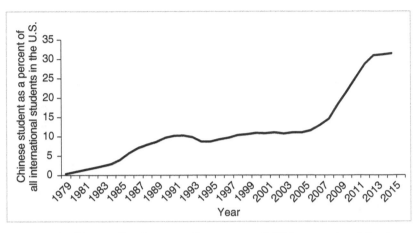

Figure 1.3 Chinese Students as a Percentage of All International Students Studying in US Institutions of Higher Education from 1979/80 to 2015/16
Source: Compiled with data retrieved from Institute of International Education (2010–16)

higher education – particularly at the BA and MA levels – has increased dramatically since the Great Recession of 2008, largely as public universities look to international students as an alternative funding source because they typically pay several times more than their in-state counterparts (Lewin 2012; Loudenback 2016).

Despite the increasing number of students leaving China to study abroad each year, the number of returning students has also increased significantly, particularly over the past decade. Preferential policies such as the various talent-attracting programs, coupled with a rapidly developing economy, are luring both recent graduates and more prominent expatriates and professionals to return (Appelbaum et al. 2016). The Chinese central government hopes that as more individuals return to China, they will bring about a cultural change to China's academic and private enterprise environment, and facilitate China's transition to a knowledge- and innovation-based economy. There are some indications that this is happening, although how much is due to returnees is unclear: a number of Chinese universities have risen in various rankings of the world's top educational institutions. Soochow University, for example, leapt from a position of 237 in 2012 to 96 in 2016 on the Nature Index publication-based scoring system,[12] and several other less well-known research institutions in China have also increased their share of high-impact research outputs (Nature 2013, 2017). At the level of basic research, however, classrooms and laboratories are not necessarily conducive to innovative thinking and scientific breakthroughs – a topic we explore in some detail in chapter 5.

Ultimately, despite ambitious five-year plans and the 15-year MLP, significant increases in governmental spending on scientific research, on initiating preference policies designed to attract world-class researchers to China, and efforts to increase the number of PhD graduates in STEM, there are some crucial underlying issues that must first be addressed if China is to become the innovation-driven economy it so desires. We explore the history of China's efforts to become an S&T leader in chapters 2 and 3, and provide some examples and case studies in chapter 5.

2 | Science and Technology in China
A Historical Overview

China's advances in science and technology long predate the creation of the People's Republic in 1949. The country's current approach to institutionalizing S&T were shaped in part by China's relationship with the West in the nineteenth century, and during the years immediately preceding Maoism. In this chapter we examine this historical legacy, focusing especially on the impact of Mao Zedong and Deng Xiaoping – the latter's revival of S&T as key to China's development, following the setbacks of Mao's Cultural Revolution. One major feature of China's historical legacy is its top-down approach to planning, a legacy that suffuses not only its economic policies but also its efforts to achieve scientific innovation and technological advances. Five-year plans and the MLP – effectively state-led industrial policies – currently guide China's S&T development. We summarize and critically review these efforts, culminating in China's current push for indigenous innovation.

THE INSTITUTIONALIZATION OF S&T IN CHINA

Science and technology develop in a broader political, social, and economic environment in China as well as elsewhere. China had maintained a relatively sophisticated level of technology until the sixteenth century. But it subsequently began to lag behind as it isolated itself from the rest of the world, especially Europe where science rapidly flourished because of, or as a result of, the Renaissance. The argument that pre-modern China was uniformly lagging in S&T development has been challenged by some historians of science, whose research seeks to show that various regions of China and disciplinary fields developed in different ways (Mullany 2017). Such differences, however, do not alter the fundamental conclusion that the introduction of modern science to China was an unintended consequence of a series of activities imposed by the West.

Alongside early capitalist expansion, foreign missionaries came to China. The missionaries not only fulfilled their task of evangelization; they also transmitted scientific knowledge, especially mathematics and astronomy, through book translations. When China was forced to open its door to Western powers after its defeat in the mid-nineteenth-century Opium Wars, some liberal-minded Qing court officials attributed China's losses to its lack of advanced technology. They campaigned for China to learn (and benefit) from the West in various ways: purchasing equipment, setting up modernized factories, creating Western-style schools, translating science books, inviting foreign experts to come to China, and – perhaps most effectively – dispatching Chinese students to study abroad.

With support from government, families, or other means, many Chinese went to study in Japan, European countries, or the US. One group of those who studied abroad stood out. These were students supported by the Boxer Indemnity Scholarship Program, established by the US and UK governments, with money from the Chinese government's forced overpayment for its alleged failure to suppress the Boxer Rebellion. The returnees, predecessors of the current sea turtles (haigui, a Chinese slang term for Chinese people that have returned to mainland China after having studied abroad for several years), were able to exert some influence on politics, economics, culture, and education. Importantly, they had a vision of transforming China by introducing modern scientific knowledge.

The returnees also brought back with them scientific societies and networks through which to promote modern science. For example, in 1914, a group of Chinese students in the natural sciences programs at Cornell University emulated the American Association for the Advancement of Science (AAAS) to form the Chinese Association for the Advancement of Science (best known as the Science Society of China), with its publication, *Kexue* (*Science*), also named after the journal of the AAAS. After moving back to China in 1918, the society made great efforts to advocate the importance of science, to persuade government and citizens to pay attention to and support scientific research, and to organize scientific research itself. While evolving into a national organization, it saw the number of its self-selected members growing from only 55 in 1914 to 3,726 in 1949. In 1922, the society founded the Nanjing Biological Survey as one of the earliest scientific research institutes in modern China. Other such institutes, including the Geological Survey, transformed from the Geological Research Institute, an institution for training geological personnel, and the Yellow Sea Chemical Engineering Institute, founded by the entrepreneur Fan Xudong.

After the 1920s, with the influx of returning Chinese students with Western doctorates in mathematics, physics, and other fields of science and engineering, a modern university system started to take shape in China. Due to their commitment and efforts, plus support from the wider society, China's undergraduate education quickly reached an international standard, and graduate education took off as well. In addition to training capable students, scientists at Peking, Jiaotong, Tsinghua, Sun Yat-sen, and Central universities were also engaged in research. When the communists seized power in 1949, there were 205 universities, of which some 30–40 were active in academic research.

Finally, the nationalist government formally put the formation of such a national academy on its agenda after its founding in 1927. The Academia Sinica was founded in Nanjing, then China's capital. Cai Yuanpei, a scholar trained in both Confucian scholarship and Western science who served as minister of education in Sun Yat-sen's government and was later first president of Peking University, was appointed as Academia Sinica's president. Under Cai's leadership, the academy attained a significant position in Chinese science during the Republican era. In 1929, another comprehensive research institution – the Peiping Academy – was established in Peiping (now Beijing). The establishment of both academies marked the beginning of China's independent system of scientific research. The Academia Sinica and the Peiping Academy were similar to research organizations in France and the former Soviet Union: functioning independently of universities, they concentrated the nation's talent by integrating research, administration, and funding for both natural and social scientists under one roof. The Academia Sinica in the south, consisting mainly of returnees from the US, had 13 research institutes in the natural and social sciences; the Peiping Academy in the north, comprised of returnees from Europe, had nine.[1] Along with scholars at other independent research institutions, there were about 700 natural scientists in China in 1949. Importantly, this system has had an enduring influence on the development of research and education in China.

SCIENCE AND TECHNOLOGY IN MAO'S CHINA: PROGRESS AND SETBACK

On November 1, 1949, exactly one month after the founding of the People's Republic, China established the Chinese Academy of Sciences (CAS) as a government agency by taking over all the research institutes under the jurisdiction of the Academia Sinica and the Peiping Academy that remained on the mainland. While the CAS gradually became the center and the driving force of scientific work for the entire nation, China also followed the Soviet model to set up research academies in various ministries. Industrial ministries set up their own specialized research institutes and laboratories at major industrial enterprises;

military-focused research institutes were also formed. Thus, the organization of Chinese science started to take shape and did not experience any dramatic changes until the late 1970s, when the Reform and Open Door policy was initiated.

In 1952, under Soviet advice, the government initiated a major reorganization of departments, colleges, and universities throughout China. The government closed all missionary universities, amalgamating them into domestic ones. It also relocated faculty and students, arbitrarily merging specialties across universities and colleges. For example, Tsinghua's colleges of arts, law, and the natural sciences became part of Peking University; its college of agriculture was merged into the Beijing Agricultural College; and Tsinghua absorbed all engineering departments from Peking University and Yenching University, a missionary institution, and became a multi-disciplinary polytechnic university. Specialty colleges were also formed. As a result of this reorganization, China's universities began to focus on education, while their role in research gradually decreased, with the gap between research and education further widening in subsequent years. The relocation of specialties also broke the internal linkage between basic research, applied research, and development, significantly impacting the training of scientific talent.

Upon taking power, the Communist Party, under Mao Zedong's leadership, was confronted with the immediate and overriding challenge of raising living standards, much of which would require focused attention by the scientific community. This in turn led to a priority for research in such areas as public health and medicine. China quickly launched campaigns to reduce mortality and to control infectious diseases; carried out work on inoculation and vaccination against smallpox, poliomyelitis, measles, rubella, and mumps; and put the prevention and treatment of diseases at the top of the list of medical services and medical research. Such efforts paid off significantly. For example, China's infant mortality was significantly reduced from 200 per 1,000 before 1949 to 9.5 per 1,000 in 2013. These effects were long-lasting: the life expectancy of the Chinese was 76.1 years in 2015, comparable to that of advanced industrialized countries.

In order to produce enough food to feed China's large population, agricultural scientists directed their efforts toward practical aspects of the science and incorporated them into the broader program of agricultural research and production. This explains why Yuan Longping, who found a new type of hybrid rice seed, became one of the first two winners of China's State Superior Science and Technology Award in 2001. Chinese scientists also envisioned scientific applications to industries from the iron and steel, crude oil, and other raw material sectors.

Meanwhile, Chinese scientists were also engaged in basic science with some impressive competency and capability. While China may have suffered in terms of numbers of top-level basic science researchers based on its size and needs, some were as good as any who could be found elsewhere. However, China's basic research all too often emphasized public appearance – work aimed at gathering world attention, but with little practical value. Politics was often key in determining which fields of basic science would become priorities. For example, China gave enormous attention to high-energy physics, one of the costliest and most demanding scientific fields, and one with the least short-term utility. On the other hand, China also gave much attention to biochemistry, a far more utilitarian field of basic science. As a result, in the 1960s, Chinese biochemists successfully synthesized bovine insulin and conducted its structural analysis, which was acclaimed as a major scientific achievement and an important indication that the Chinese scientific effort was about to achieve quality in a growing number of fields; the achievement was even nominated for a Nobel Prize in 1980. Research in biochemistry continued during the Cultural Revolution, a time when almost all scientific research activities were severely disrupted. And on the ideological front, research in biochemistry was also interpreted as confirming hypotheses that Friedrich Engels had made in *The Dialectics of Nature*.

Another impressive achievement was the discovery that artemisinin, an extract from a traditional Chinese herb, is effective in treating malaria – a discovery with political implications. In 1964, during the Vietnam War, Ho Chi Minh, the leader of North Vietnam asked Mao Zedong for help in developing a malaria treatment for his soldiers. Because malaria was also a major public health threat in China's southern provinces, Mao agreed to set up a secret drug discovery project, named Project 523 after its starting date, May 23, 1967. At one time, the project mobilized medical personnel from seven provinces, who spent a couple of years collecting and screening more than 3,000 traditional Chinese herbs for malaria-curing ingredients. In 1969, Tu Youyou, then a junior researcher at the Chinese Academy of Traditional Chinese Medicines, joined the project, searching ancient literature on traditional Chinese medicine for clues to develop malaria therapies. The plant *artemisia annua* turned out to be a potentially viable candidate, and Tu then came up with a novel purification procedure, using low-temperature ether to extract the effective compound and thus making the breakthrough. Though the team published a joint paper without listing the authors, Tu's contribution to the discovery was recognized in 2015 by a Nobel Prize in Psychology or Medicine – a significant accolade as it was also mainland China's first Nobel Prize in science (Rao et al. 2015).

Although there were some breakthroughs in biochemistry and a few other areas, efforts in Mao's China were primarily shaped by multi-year S&T plans, with scientists carrying out major programs to meet the demands of a politically and ideologically new environment that might also enhance their careers and professional development. The first plan of its kind was the 12-year (1956-67) S&T plan. The plan prioritized five fields for investment: atomic energy, electronics, jet propulsion, automation, and rare mineral exploration – all directly applicable to the development and subsequent delivery of nuclear weapons. A secret national defense S&T plan was also distributed with the 12-year plan, which clearly listed atomic energy, jet propulsion and rocketry, semiconductors, computers, and automation as top priorities. For that purpose, the government in the 1950s set up seven ministries of machine building, targeting research, development, and production activities in industries such as atomic energy, shipbuilding, electronics, aeronautics, and aerospace with military applications.

Within seven years, China successfully developed, refined, and deployed nuclear bombs and strategic missiles, and successfully launched man-made satellites (*liangdan yixing*), with relatively little foreign assistance. Such achievements made the political point that China had started to be on a par with other great powers. The 12-year plan for the first time advocated a so-called "missions-driven disciplinary development" strategy, with missions having apparent political orientations. The subsequent development of Chinese science continues to reflect this early planning legacy – politicization of science, state-led research endeavors, big science, the concentration of resources, and top-down involvement and interference in S&T enterprise development.

One result of Beijing's top-down intervention in S&T was tension between scientists (and, more generally, intellectuals) and the party-state during the first 30 years of the PRC. Initially, intellectuals did not perceive what their work and life would be like under communist control. University professors tried to retain autonomy in their professional work, only to find that they had to be relocated to a different and often unfamiliar institution. Indeed, the reorganization of departments, colleges, and universities mentioned above deprived many of the pre-1949 intellectuals of their institutional shelter, where they were able to enjoy certain academic freedoms by exercising autonomy in teaching and research, maintaining some degree of authority over their own institutions, and pursuing their own interests. Intellectuals started to feel unconformable with the CCP, as they were often required to engage in sessions of ideological education and self-criticism; at times they also became targets of political campaigns. Many of them were badly treated in the Anti-Rightist Campaign in 1957, with more than half a million of them, including a significant number of high-ranking scientists, persecuted as "rightists." After the Anti-Rightist Campaign, even the survivors lived in a perpetual state of fear, afraid to speak out, and often unable to work in their chosen fields while forced to attend endless political-study sessions.

But the CCP soon realized that its leadership over Chinese science could only be achieved through the endeavors of scientists themselves, since very few party members were knowledgeable in modern science. Therefore, the CCP felt it had no choice but to depend upon nationalist- and foreign-trained scientists to build China's strategic weapons. In the aftermath of the Anti-Rightist Campaign – and especially the withdrawal of Soviet support, which made it more urgent to emphasize indigenous efforts – the CCP modified its hostile policy, at least toward scientists (other members of the intellectual community were not always so fortunate). Out of concern that the scientists' expertise would be required in socialist construction and the completion of the 12-year plan, Mao suggested removing the "rightist" label from most of the rightists in the first amnesty on the eve of the 10th anniversary of the founding of the PRC. The leadership also recalled scientists to key positions in research and education, and even allowed former "rightists" to return to their old positions.

The prevailing anti-intellectual trend, and the resulting dissatisfaction on the part of many scientists and other intellectuals over what they perceived as their unfair treatment, led the CCP to address these issues at its second conference on the topic (the first conference on the issue of intellectuals was held in 1956 to prepare for the formulation of the 12-year S&T plan). At the conference held in Guangzhou in February and March 1962, which was originally set for planning science, Premier Zhou Enlai reassured intellectuals that they "work enthusiastically for socialism, accept the party leadership, and are ready to go on remolding themselves so that they now belong to the working class and should no longer be regarded as bourgeois." Thus, they should be given "due confidence and support." Vice Premier Chen Yi boldly declared, "China needs intellectuals, needs scientists. For all these years, they have been unfairly treated. They should be restored to the position they deserve." Chen Yi even encouraged intellectuals to take off the hat of "bourgeois intellectuals" and put on the crown of "intellectuals of the working people" (Nie 1988).

Both speeches gave Chinese intellectuals new hope that their services were finally needed and their contributions appreciated. Then, the "Fourteen Articles on Scientific Work," also known as the "Constitution of Science," which the party formulated in 1961, crystallized the CCP's new policy as uniting with, educating, and remolding intellectuals. In particular, the CCP wanted to utilize the expertise of intellectuals while at the same time keeping a cautious eye on them, especially those of the pre-1949 and "black" class background,[2] and to engage them in political education, including the study of communist ideology and Marxism-Leninism-Mao Zedong Thought. Ultimately, the party would repudiate intellectuals' call for autonomy in their professions, instead of seeking to transform them into "red experts" with socialist and communist consciousness (Nie 1988).

In sum, despite an overall anti-intellectual environment that lingered for most of the period between the late 1950s and 1966, the administrators involved in the strategic weapons programs, who had extensive military experience but limited (or no) modern scientific knowledge, were compelled to trust scientists. Such trust was furthered by enrolling some scientists into the party, both as a way of mobilizing their support and as a form of reconciliation. Scientists working on the strategic weapons programs were also granted a significant amount of autonomy. Leading figures, such as Qian Sanqiang of the atomic bomb program and Qian Xuesen of the missile program, were given direct access to central leadership, including Mao himself, so that they were able to make independent scientific decisions (Feigenbaum 2003).

These developments led to a far less politicized academic environment, where scientists were able to spend more time on research and teaching instead of studying political ideology and engaging in self-criticism.

However, this honeymoon period that Chinese intellectuals enjoyed with the party proved to be short-lived: a bigger political storm was fast approaching. In 1966, in his fight with his rivals within the CCP leadership, Mao Zedong also mobilized students and junior intellectuals to attack "bourgeois, reactionary academic authorities," namely, high-ranking intellectuals, who were criticized and ruthlessly attacked for aiding and abetting so-called "capitalist roaders." Consequently, the Cultural Revolution decade 1966–76 became a nightmare for Chinese intellectuals, who, as a social group, were denounced as "stinking number nine" (*chou laojiu*) at the bottom of the barrel as social outcasts after landlords, rich peasants, counter-revolutionaries, bad elements, rightists, traitors, spies, and capitalist roaders. Their "stinkiness" was not a stigma of individual wrongdoing, nor a lack of faith in the party, nor an unwillingness to sacrifice oneself for socialism, but was associated with their "bad" class inheritance.

Under an overall environment in which the CCP distrusted and even became suspicious of intellectuals, the development of the strategic weapons programs was an exception. Although some scientists were spared the worst of the turmoil and even protected early on, they were not completely safe and also saw their work disrupted. For example, Qian Sanqiang, the leader of the atomic bomb program, was marginalized long before he was sent to the countryside immediately after the bomb's first successful detonation in 1964. Some scientists were attacked in big-character wall posters (*dazibao*), criticized and humiliated at public meetings, and investigated and interrogated by radicals. Their homes were searched and their properties confiscated, and they were abused and tortured physically and psychologically. University professors were accused of "poisoning young students" through their teachings. If they studied abroad, they were labeled American or Soviet "spies." Many were deprived of the right to teach and carry out research; they became subjects of virulent attacks and were dismissed from their jobs and sent to the countryside or frontier areas for "labor reform" (Cao 2013). Several senior scientists who had played an active role in the strategic weapons programs lost their lives during these tumultuous 10 years, although they were honored posthumously by the state for their contributions.

S&T UNDER DENG XIAOPING: REFORM AND CATCHING UP

Soon after the end of the Cultural Revolution, the political leadership under Deng Xiaoping started to revive China's S&T system and rehabilitate intellectuals, including scientists. With the call to modernize agriculture, industry, national defense, science and technology, universities reopened their doors to students admitted through the national higher education examination (*gaokao*), a practice that was disrupted in the early period of the Cultural Revolution. Universities also started to integrate research into student training whereby academics became active at the frontier of international science.

Moreover, amid an overall Reform and Open Door agenda initiated in the late 1970s, as well as a realization from the experience of other countries that scientific research contributes to economic growth, it became apparent to China's leadership that the lack of linkage between research and the economy was a serious problem that needed to be solved. Indeed, following the Soviet model, China's R&D activities had for some three decades largely been confined to institutions of learning where scientists were more interested in producing academic publications than producing prototypes. As such, most research results did not benefit the nation's economic development and ended at one of the three stages – samples, exhibits, and gifts – without reaching the stage of marketable commodity (Lu 2000).

In 1982, the CCP called for the reliance of economic development on science and technology. Three years later, the CCP's Central Committee and State Council's *Decision on the Reforms of the S&T System* set fundamental objectives of the reform to broadly and rapidly apply results from research to production. Thereafter, China's science, technology and innovation policy agenda has evolved from acknowledging S&T as a productive force to the reform of the S&T system, then to the initiation of the strategy of "revitalizing the nation through science and education;" from the construction of a national innovation system (NIS) to the pursuit of indigenous innovation, then to improve the NIS itself (Liu et al. 2011).

During the early period of the reform, the CAS implemented a "one academy, two systems" experiment, keeping a small number of its research personnel in basic research while pushing the rest to seek outside support for applied R&D that was seen as directly benefiting the economy and meeting market needs. High-tech companies started to spin off from CAS research institutes as well as Peking and Tsinghua universities and others in Beijing's Zhongguancun area, the nation's most talent-intensive region. In fact, technological findings from the CAS were turned into marketable products by more than one-third of the firms along the so-called "Electronics Street" in Zhongguancun. These enterprises focused on meeting market needs and integrated technological development, industry, and commerce. Operating on the principle of "self-chosen partnership, self-financing, self-operation, and self-responsibility for gains and losses" (the four-"self" principle), they promoted technology transfer from universities and research institutes and created a series of products with market potential and competitive edge. Some of them were listed at home and abroad. As a whole, in its initial development, Zhongguancun saw a celebration of S&T spin-offs and startups not previously witnessed in China's history. Entrepreneurial activities also spread to other Chinese cities, and national S&T parks were established in Zhongguancun and elsewhere in China to pursue high-tech commercialization (we discuss one such park, Suzhou Industrial Park, in chapter 5).

As the country became more open, especially after it was accepted into the World Trade Organization in December 2001, China has raised its international profile through dispatching an increasing number of students abroad to experience and to be involved first-hand in the advancements of scientific research and development in new and high technology, strengthening its cross-border S&T cooperation, and receiving technology and knowledge embedded in foreign direct investment (FDI). Multinational companies (MNCs) have not only increased their investment in China but also turned their China operations into critical components of their global high-end networks of innovation. While quite a number of foreign corporate R&D centers are engaged in localizing technologies developed elsewhere, Microsoft, Intel, and other MNCs have gained tremendously by investing in R&D in China as first-movers. Meanwhile, China has also witnessed an increasing return of its expatriates from overseas stays for study and research, some of whom have given up lucrative permanent positions abroad for professional opportunities in their motherland. As we discussed in the previous chapter, returnees bring back with them knowledge, experience, and skills as well as connections, thus facilitating cross-border technology transfer and helping stimulate high-tech entrepreneurship (for further discussion of the role of returnees, see chapters 3 and 5).

OVERVIEW OF PROGRAMS, PLANS, AND PRIORITIES

China's pattern in research organization does not completely disappear in the Reform and Open Door era. Planning and the top-down approach are still apparent in developing S&T plans and R&D programs and setting up priorities for research activities, while international practice in the assessment of research and achievements have been introduced.

The establishment of a national funding agency in science, technology, and innovation represents one such example.

Historically, China did not have sufficient resources to fully fund research activities, even important ones, to achieve its ambition. Consequently, for quite some time, the state had to selectively support research projects that were regarded as especially important, most being applied research and related to national defense. Other than that, it did not actually distinguish between various levels or types of institutions or research activities, but furnished a lump sum uniformly from higher-level administrative departments to the work unit (*danwei*) as operating expenses according to headcounts. Because the state could not exert strict control to ensure that funds were used in the most constructive and efficient manner, working on easy but less important projects and duplicating others or foreign research were popular. This funding model had not changed until the reform of the S&T system in the mid-1980s, which required that most of the applied research institutes be market oriented and financially self-sustainable. Apparently, at stake was basic research that would have long-term impacts on the nation's scientific well-being.

Under these circumstances, in May 1981, 89 senior scientists affiliated with the CAS Academic Divisions proposed to the CCP's Central Committee and the State Council that a science foundation be established within the CAS to support basic research. The proposal was promptly approved, and the CAS Science Foundation was set up in the following year with 30 million yuan (US$15.9 million) specially allocated from the state budget. The foundation, attached to the CAS in its administration, was supervised by an independent committee of researchers from the sciences, education, industry, agriculture, medical research, and public health, thus serving the function of a national, instead of an institutional, funding agency for research.

The successful experiment of the CAS Science Foundation paved the way for the reform of the nation's research management and the

1985 decision on S&T reform proposed the establishment of a science funding system. In February 1986, the State Council decided to evolve the CAS Science Foundation into the National Natural Science Foundation of China (NSFC). Compared with the situation where the party-state determined research priorities, the establishment of the NSFC indicates that Chinese scientists were becoming increasingly more involved in decision making related to research funding. Modeled on the US National Science Foundation (NSF) and following the principle of "relying on experts, practicing democracy, supporting the best, and being both fair and reasonable," the NSFC mainly supports three categories of research projects – general, key, and major – based on project importance (NSFC 2018). In 2016, the NSFC allocated a total of 27 billion yuan (US$4.1 billion) to various projects, representing a significant increase over its first year of funding at 80 million yuan (US$23.2 million).

One of the functions of the national S&T programs historically was to achieve quality and enhance performance through setting priorities and concentrating resources. This was reflected in the principles of "anchoring one end, freeing up the other" and "doing what we need and attempting nothing where we do not," which redefined the role of the state regarding research support. Priority setting would be both top-down and bottom-up, insofar as some of the priorities were politically driven rather than purely scientific. In recent years, this has also involved a radical reduction in the traditional role of supporting applied R&D in industry, as well as a more concentrated focus on areas of science most susceptible to "market failures" – basic research, education, pre-competitive R&D, and research for public missions. Notable programs during the Reform and Open Door era included the State High-Tech Research and Development Program and the State Basic Research and Development Program.

In March 1986, amid the global high-tech revolution, four senior CAS scientists who had been involved in China's strategic weapons program – Wang Daheng, Wang Ganchang, Yang Jiachi, and Chen Fangyun – wrote to Deng Xiaoping, then China's paramount leader, suggesting following the world's high-tech trends and developing China's high technology. The suggestion was immediately approved, and a State High-Tech Research and Development Program, was outlined. Mostly known as the 863 Program to commemorate that the program was initiated in March 1986, it was administered by the Ministry of Science and Technology (MOST) with an initial allocation of 10 billion yuan (US$2.9 billion) and a significant emphasis on seven fields – biotechnology, space technology, information, lasers, automation, energy, and new materials.

Although the 863 Program also included basic research components underpinning high technologies, focus was placed on achieving technological advantages. Therefore, a program to predominantly target basic science and mission-oriented fundamental research was initiated in 1989 to make up the gap in research. The so-called "Climbing Program" intended to attract China's brightest scientists to work on 30 critical research projects, such as high-temperature superconductivity, non-linear science, important chemical problems in life processes, and brain function and its cell and molecular basis. This program evolved into a State Basic Research and Development Program, or 973 Program, to commemorate the initiation of the program in March 1997, which would showcase China's endeavors in pursuing excellence in basic research. Also under the administration of MOST, the program received an initial investment of some 2.5 billion yuan (US$302 million) over five years (1998–2002) to support projects at an average level of 30 million yuan (US$3.6 million). The projects to be supported were in six broad areas relevant to the nation's economic and social development – population and health, information, agriculture, resources and the environment, energy, and new materials. Projects also were required to meet one of three criteria: first, they attempted to solve major problems associated with China's social, economic, and scientific and technological development; second, they were related to major basic research problems with interdisciplinary and comprehensive significance; and third, they could take advantage of China's strength and special characteristics such as natural, geographic, and human resources and help China occupy "a seat" at the frontier of international research.

Most of these national S&T programs have disappeared as the most recent round of reform has attempted to integrate all of them into five streams. These streams include basic research,[3] major national S&T programs,[4] key national R&D programs,[5] special funds to guide technological innovation,[6] and special funds for programs for developing human resources and infrastructure.[7]

In addition to national S&T programs, there have been other ministerial, institutional, and organizational programs targeting specific areas with respective jurisdictions. For example, the Minister of Education (MOE) launched the 211 Program in 1995 and subsequently the 985 Program in 1999, aimed at turning some Chinese universities into world-class institutions. A significant amount of public funds has been invested in the improvement of the infrastructure at Chinese universities, especially leading ones, and setting up "key" national laboratories open to both domestic scientists and those from abroad has become a common practice (Simon and Cao 2009b). The MOE is now launching another program, Developing World-Class Universities and First-Class Disciplines, also known as World Class 2.0, continuing its push whereby some Chinese universities or disciplines could be at the international frontier of excellence in research and teaching (Q. Wang 2017).

The CAS has also experienced some significant changes to its organizations, priorities, and operation. As mentioned, at the height of the S&T management system reform in the mid-1980s, the academy implemented a "one academy, two systems" policy by concentrating most of its efforts on research that directly benefits the Chinese economy and spinning off enterprises themselves to meet the needs of the market, while at the same time continuing with basic research. Such a move indicates a substantial departure from the academy's tradition of being a purely academic institution. Starting from 1998, the CAS launched a Knowledge Innovation Program, aiming to build CAS into the nation's knowledge innovation center in the natural sciences and high technology, and a base for world-class state-of-the-art scientific research, fostering first-rate scientific talent and for the development of high-tech industries (Suttmeier et al. 2006a; 2006b). And most recently, the CAS launched another program, Pioneering Action Initiative, to orient the academy toward major national demands, the international frontier of science, and the major battleground for the national economy (Cao and Suttmeier 2017).

Finally, as noted previously, China has retained its strong belief in planning S&T development, continuing this practice in the Reform and Open Door era. In early 2006, the government launched the Medium to Long-Term Plan for the Development of Science and Technology (2006–20) mentioned above and in chapter 1, showing the ambition to turn China into an innovative nation by 2020. The MLP specified the following quantitative objectives for the year 2020, including:

* investing 2.5 percent of its increasing gross domestic product (GDP) in R&D;
* raising the contributions to economic growth from technological advances to more than 60 percent;
* limiting its dependence on imported technology to no more than 30 percent; and
* becoming one of the top five countries in the world in the number of invention patents granted to Chinese citizens and in the number of citations to Chinese-authored scientific papers.

The plan consists of a number of components. The first sets out guidelines and general principles derived from the objectives of having science and technology lead future economic development, enabling China during the plan period to "leapfrog" into positions of leadership in emerging science-based industries, and establishing a capability for "indigenous innovation" (*zizhu chuangxin*, also translated as "independent" or "homegrown" innovation). In particular, the MLP points to *zizhu chuangxin* as having three components: genuinely "original innovation," "integrated innovation" (or the fusing together of existing technologies in new ways), and "re-innovation," which involves the assimilation and improvement of imported technologies.

A second part of the plan identifies its priority areas and programs. They include 11 broad "key areas" pertaining to national needs and eight areas of "frontier technology." Within these, the plan identifies a series of priority projects. For instance, under the "new materials" area, the plan includes work on smart materials, high-temperature superconducting technology, and energy-efficient materials.

In addition to the priority areas, the MLP identifies a series of large national "mega-programs" in engineering and science. In the area of basic research, the MLP includes foci on the development of new disciplines and interdisciplinary areas, science frontiers, and fundamental research in support of major national strategies, as well as the mega-programs such as protein science, quantum research, nanotechnology, and development and reproductive biology, with climate change and stem cell programs being added later.

The final sections of the plan, as well as an accompanying document, deal with the introduction of a policy framework for the plan's implementation, including preferential tax policies, policies for high technology industry zones, the assimilation of foreign technology, and the strengthening of intellectual property protection for Chinese innovators. They also include important policies to strengthen and diversify funding for science and technology, make expenditures more efficient, and develop the nation's human resources for science and technology, including the cultivation of world-class senior experts, an expanded role for scientists and engineers in industry, policies to recruit talent from abroad, and reforms in education to support the goals of greater creativity and innovation (Cao et al. 2006).

SOME CHALLENGES FACING CHINA'S S&T AND INNOVATION SYSTEM

Despite – or perhaps because of – its state-led efforts, China's S&T system faces some major challenges. Indeed, despite rising gross expenditure on R&D, which reached 2.1 percent of China's GDP in 2016, more and better-trained researchers, and sophisticated equipment, Chinese scientists have yet to produce cutting-edge breakthroughs worthy of a Nobel Prize in science.[8] Few research results have been turned into innovative and competitive technology and products. With few exceptions, Chinese enterprises still depend on foreign sources for core technologies. These challenges are also fundamentally associated with the deep-seated structural and institutional problems that have existed in China's S&T system – lack of inter-governmental coordination of activities at the macro level, inefficiency in funding distribution at the meso level, and inappropriate performance evaluation of individuals and institutions at the micro level (Cao et al. 2013a).

Meanwhile, the Chinese government missed the opportunity of the post-2008 global financial crisis to deepen its reforms; in fact, the Chinese response to the crisis delayed much-needed reforms, including reform of the S&T system. Having maintained average annual GDP growth of about 9 percent between 2008 and 2013, China's economy seems to have reached a plateau, or a "new normal," characterized by much slower growth: as we have previously noted, China's GDP growth was 6.8 percent in 2017, roughly where it appears to have stabilized in recent years. Additionally, the "new normal" is predicted to last for a prolonged period of time (Wen et al. 2016). China's new political leadership, under Xi Jinping, understands that it not only inherits the legacy of an economy that had been growing at almost 10 percent on average for the past decade but also faces challenges, ranging from inclusive, harmonious, and green development to an aging society and the "middle income trap." These challenges, coupled with the "new normal," further highlight the urgency and necessity for China to transform its economic development model from one that is labor-, investment-, energy-, and resource-intensive into one that is increasingly dependent upon technology and innovation.

In the next chapter, we examine the policy framework that shapes China's approach to S&T innovation, in order to better understand both the strengths and weaknesses of China's top-down approach.

3

China's Science and Technology Enterprise
Can Government-Led Efforts Successfully Spur Innovation?

Thanks to policies, programs, and priorities, reviewed in the previous chapter, China has begun to see the emergence and then the improvement of a national innovation system. This has had a significant impact on R&D efforts at the national, provincial, and local levels. Universities and public research institutes have taken steps to provide direct economic benefit through spinning off high-tech enterprises with promising products and technologies. In recent years, high-tech startups have played an increasingly significant role in indigenous innovation. Returnees also have contributed to the acceleration of China's innovation endeavor. Along with these efforts, national defense-oriented R&D organizations, as well as universities and private enterprises, have also started to develop dual-use technologies. The most recent reform of public R&D funding has changed the funding structure not only at the government level but most significantly at enterprises that have to invest their own money in innovation if they hope to compete domestically, and increasingly, internationally (Cao and Suttmeier 2017).

A key focus of this chapter is to evaluate the role of the Chinese government as a driver of S&T development in such key fields as life sciences, nanotechnology, advanced manufacturing, aerospace, clean energy, and supercomputing. The CCP continues to control scientific enterprise through a Leading Group on Science, Technology, and Education at the State Council. The central government has initiated various top-down national programs and has approved the setting up of 168 national high-tech parks. The central government still controls a significant share of the R&D funding, which is often distributed without necessary regard to merit or transparency; and its use has all too often been ineffective, inefficient, wasted, and abused (Cao et al. 2013a; Sun and Cao 2014; Cao and Suttmeier 2017). Therefore, it is necessary to take a critical look at state-led efforts, including the growing emphasis on S&T parks as innovation hubs.

LEADERSHIP'S COMMITMENT

There is little doubt that China would not have achieved its current global status in science, technology, and innovation without the commitment and support of its leadership, a continuity that dates back to Deng Xiaoping, China's paramount leader in the early period of the Reform and Open Door era. Deng's successors, from Jiang Zemin, Hu Jintao, to Xi Jinping, all have championed the role of technology and innovation in China's continuous economic development.

For example, in May 1995, a National Conference on Science and Technology was held under Jiang Zemin's leadership (a similar conference in 1978 under Deng Xiaoping led to the so-called "spring of science"). In preparing for the conference, the CCP's Central Committee and China's State Council, the two highest party and state organizations, issued the *Decision on Accelerating the Progress of Science and Technology* to echo an earlier similar document by the organizations, the *Decision on Reforming the Science and Technology Management System*. Indeed, the conference and the decision in 1995 proposed for the first time a now well-known strategy – "revitalizing the nation through science, technology, and education" (*kejiao xingguo*). In particular, the strategy set up the principles of China's science and education development toward the twenty-first century by defining the role of S&T as "orientation, dependence, and climbing up new heights." That is, S&T must be oriented toward the economy, economic development must depend upon S&T, and S&T must strive to achieve a global status. It stressed the importance of raising the nation's indigenous innovation (*zizhu chuangxin*) capability for the first time and the importance of technological development at the enterprise level. High technology was to be given enormous attention in the national economy and the nation's industrial policy, while basic research would shoulder the responsibility of pushing forward future economic and social development. The new strategy also stipulated an increase in investment in science and education by requiring that China's gross R&D and education expenditures reach 1.5 percent and 4.0 percent of the gross domestic product respectively by the year 2000. This, along with another strategy of "empowering the nation with talent" (*rencai qiangguo*), put forward in the early twenty-first century, ushered in a new era for technology and innovation in China (Cao 2002).

The *kejiao xingguo* strategy also anticipated the indigenous innovation strategy formally introduced some 10 years later when the state unveiled the 15-year Medium and Long-Term Plan for the Development of Science and Technology (2006–20). Of course, the MLP further raised the bar as it called for China to become an "innovation-oriented society" by the year 2020, and a world leader in science and technology by 2050 (details of the Plan are outlined in chapter 2). The plan is likely to have important impacts on the trajectory of Chinese development and thus warrants careful attention from the international community (Cao et al. 2006; 2009). These goals, and others, are provided in China's *13th Five-Year Plan for Science, Technology, and Innovation* (2016–20), discussed in chapter 1.

Organizationally, China set up a leading group for science and technology within the State Council in 1983 when Zhao Ziyang was the premier. In March 1996, after the National S&T Conference, it evolved into the State Leading Group for Science, Technology, and Education under Premier Zhu Rongji, with education added to show the government's new emphasis. The leading group, always composed of the chiefs of the leading science, education, and economic bureaucracies, is responsible for studying and reviewing the nation's strategy and key policies for the development of science, technology, and education; for discussing and reviewing major tasks and programs related to science, technology, and education; and for coordinating important issues of science and education involving different agencies under the State Council and different regions (Liu et al. 2011; see figure 3.1).

XI JINPING ADMINISTRATION'S ATTENTION TO INNOVATION

The political leadership's attention to science, technology, and innovation in post-1978 China also is at least in part the result of the fact that many of the leadership were trained in science and engineering at some of the nation's top universities (Suttmeier 2007). Only relatively recently, in its current top leadership, did China start to see the arrival of social scientists: Xi Jinping, China's President and General Secretary of the CCP's Central Committee, holds a PhD in law from Tsinghua University (his undergraduate education, as a "worker-peasant-soldier" student during the Cultural Revolution, was nonetheless in chemical engineering at Tsinghua); and Li Keqiang, premier of the State Council, obtained his undergraduate degree in law and his PhD in economics from Peking University. However, the change in educational background of the leadership does not mean that attitudes toward science and technology have changed among the top leaders.

On July 17, 2013, soon after taking over the reins of the party and state, Xi Jinping visited the Chinese Academy of Sciences (CAS), the nation's leading institution for science and research. His articulation of the problems facing S&T development in China was distilled into "four mismatches" (*sige buxiang shiying*): between the level of technological development and the requirements of socioeconomic development; between the S&T system and the requirements of science and

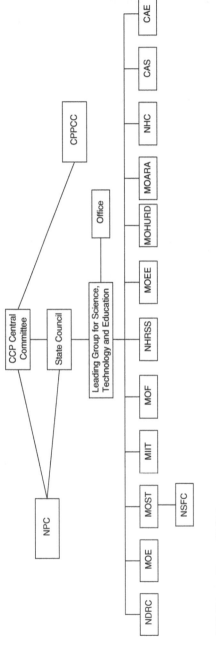

CCP: Chinese Communist Party

NPC: National People's Congress

CPPCC: Chinese People's Political Consultative Conference

NDRC: National Development and Reform Commission

MOE: Ministry of Education

MOST: Ministry of Science and Technology

MIIT: Ministry of Industry and Information Technology

MOF: Ministry of Finance

NHRSS: Ministry of Human Resources and Social Security

MOEE: Ministry of Ecological Environment

MOHURD: Ministry of Housing and Urban-Rural Development

MOARA: Ministry of Agriculture and Rural Affairs

NHC: National Health Commission

CAS: Chinese Academy of Sciences

CAE: Chinese Academy of Engineering

NSFC: National Natural Science Foundation of China

Figure 3.1 China's S&T Policy-Making Framework

technology for the system to develop rapidly; between the distribution of S&T disciplines and the requirements of science and technology for these disciplines to develop; and between existing S&T personnel and the requirements of the nation in terms of talent. Xi further stressed that the CAS should strive to become the country's main source of innovative ideas, talents, and achievements by virtue of its unique position as an organization with triple roles as a research institution, a national academy, and an educational institution. Xi particularly urged the CAS to become "a pioneer in four areas" (*sige shuaixian*): a pioneer in leapfrogging to the forefront of international science and technology; in enhancing the nation's innovative talent pool; in establishing the nation's high-level think tank in science and technology; and in becoming a world-class research institution (Poo and Wang 2014).

China's recent political leadership also is enthusiastic about broadening knowledge among its members. This is illustrated by the fact that since 2002 the Politburo of the CCP's Central Committee, or China's *de facto* governing body, has held frequent group-study sessions by inviting leading Chinese scholars to lecture on subjects related to China's socioeconomic development, many of which are technology- and innovation-focused. The Xi leadership has followed such a tradition. On September 30, 2013, for example, the Politburo held a group study at Beijing's Zhongguancun Science Park. During this ninth group-study session of the new leadership and the first ever held outside the CCP's Zhongnanhai headquarters, members of the Politburo showed particular interest in new technologies such as 3D printing, big data and cloud computing, nanomaterials, biochips, and quantum communications. In his speech on the occasion, Xi Jinping stressed that China should focus on integrating innovation with socioeconomic development, enhancing the capability for indigenous innovation, nurturing talent, constructing a favorable policy environment for innovation, and continuing to open up and engage in international cooperation in science and technology (Xinhua 2013).

Meanwhile, the new leadership is particularly keen on the issue of weaving together the so-called "two layers of skin" (*liang zhang pi*), research and the economy, a long-lasting challenge for China's S&T system that the post-1985 reform has tried to address. The main topic of discussion at the seventh meeting of the Central Leading Group for Financial and Economic Affairs on August 18, 2014, chaired by Xi, for example, was a draft document calling for deepening institutional and system reform and accelerating an innovation-driven development strategy. The CCP's Central Committee and China's State Council endorsed the strategy on March 13, 2015, and formally released an outline of the strategy a year later. The 13th Five-Year Plan (2016–20), discussed at the fifth plenum of the 18th CCP's Central Committee in 2015 and passed at the fourth session of the 12th National People's

---

Box 2. Intelligent Manufacturing and Robotics

As we have noted, thanks largely to its large supply of cheap labor, for several decades China has been seen as the world's factory. As China attempts to move up the global value chain and transition to a knowledge-based economy, the central government knows that it can no longer depend on low-end manufacturing as the country's primary driver for economic growth. In addition, manufacturing wages have been rising steadily, from an average annual income of approximately 21,000 yuan (~US$2,800) in 2007 to approximately 59,000 yuan (~US$8,900) in 2016 (Trading Economics 2017a). In an effort to overcome rising wages and salaries, and increasingly toxic environmental pollution, China has set its sights on intelligent manufacturing and robotics. China's leadership is no longer satisfied with manufacturing low-end products such as apparel and electronics; it aims to reduce its dependence on foreign imports of core components by producing them domestically. For example, China is almost completely dependent on foreign production of high-level digital control systems and high-level hydraulic components (Hsu 2017).

In 2015, China launched its US$1.5 billion, 10-year roadmap called *Made in China 2025* (MIC 2025) to overhaul and upgrade the country's manufacturing capabilities (State Council 2017a). Many countries, however, including the US and the EU, have expressed concerns that MIC 2025 appears to provide domestic companies with preferential subsidies and market access to promote domestic innovation and increase overall competitiveness (European Union Chamber of Commerce in China 2017; US Chamber of Commerce 2017). Despite these concerns, China is forging ahead with its MIC 2025 strategy for advanced manufacturing and robotics. China acquired 75,000 industrial robots (i.e. automated manufacturing machines) to become the world's largest importer of robots in 2015, and installed an additional 90,000 robots in 2016, accounting for one-third of the world's total (Knight 2015; Lawton 2017). Despite the increase, China's robot density (defined by the International Federation of Robotics as the number of multipurpose industrial robots per 10,000 persons employed in the manufacturing sector) is well below that of developed countries, indicating that the point of saturation is still very far off and there is a huge potential for future growth (International Federation of Robotics 2016).

Like other countries with advanced manufacturing, industrial robots are expected to displace an increasing number of manufacturing workers, something that has already taken place across China. For example, the Guangdong Janus Precision Manufacturing plant acquired 160 industrial robots in 2016 and decreased the number of human employees from 197 to just 33 (State Council 2017b). As China's manufacturing sector becomes more automated, the need for workers will transition from low-waged migrant workers to skilled technicians who have the knowledge and know-how to control the advanced machineries.

Congress (NPC) in 2016, sets innovation as core to China's next round of development along with the other major concepts of coordination, green, opening up, and sharing (Xinhua 2016). And the five-year plan lists a number of strategic technologies and research domains intended for achieving major breakthroughs by 2030, from aero-engines, quantum communications, intelligent manufacturing and robotics, deep-space and deep-sea exploration, major new materials, to neurosciences (Ministry of Science and Technology 2016). All these demonstrate the importance that the new leadership has attached to innovation for the restructuring of China's current economic development model.

Realizing the urgency of making innovation contribute more to the economy as well, along with overcoming the obstacles that hinder China's socioeconomic development to the next stage, the leadership also has pushed forward the reform of China's S&T system. The most recent round of reform got under way in early July 2012, when the National Conference on Science, Technology, and Innovation was convened shortly before the leadership transition. Later that month, the newly established State Leading Group of Science and Technology System Reform and Innovation System Construction convened its first meeting. Made up of representatives from 26 government agencies and headed by Liu Yandong, a member of the Politburo of the CCP's Central Committee and state councilor, the leading group is mandated to guide and coordinate the reform and construction of China's national innovation system, in addition to discussing and approving key regulations.[1] When the country's top leadership changed in late 2012 and early 2013, Liu not only kept her party position but also was promoted to vice premier within the State Council, thereby ensuring continuity in her oversight of the nation's science and education portfolio and confirming the importance attached to technology and innovation. One key outcome of the conference was an official document, *Opinions on Deepening the Reform of the Science and Technology System and Accelerating the Construction of the National Innovation System*, released in

September 2012, again, by the CCP's Central Committee and the State Council.

The reform of the S&T system has accelerated since the change in political leadership. In general, the entire reform program under the Xi administration is characterized by the so-called "top-level design" (*dingceng sheji*). This entails strategic considerations in formulating guidelines so as to ensure that the reform is comprehensive, coordinated, and sustainable; a balanced and focused approach toward reform that takes into consideration the interests of the party and state; and a focus on overcoming institutional and structural barriers, as well as deep-seated contradictions, while promoting coordinated innovation in economic, political, cultural, social, and other institutions. The "top-level design" also has been exercised in the reform of the S&T system, which has strong political backing, with Xi's visit to the CAS and Politburo's Zhongguancun group study setting the course. He also has been very hands-on in reforming China's elite membership system at the CAS and the Chinese Academy of Engineering, the broader reform of the CAS, and the reform of funding mechanisms for the centrally financed national S&T programs.

Xi took time to preside over the presentation of reports by the relevant government agencies on progress with the reform of the S&T system and, most importantly, the formulation of an innovation-driven development strategy, and he has discussed and spoken about innovation on numerous occasions (Literature Research Office of the CCP's Central Committee 2016). When it was finally released in May 2016, the innovation-driven development strategy became the directive for how innovation is to be integrated into Chinese policy across the economy (CCP's Central Committee and State Council 2016). It contains the first reference to the mandating of the integration of civilian and military technology development, which is now driving much of China's innovation in artificial intelligence, information and communications technologies, nanotechnology, and many other fields. Such an emphasis, which clearly has its roots in US practice, where civilians have been developing "spin-on" military technology (see, for example, Weiss 2014), also has come up in other policy documents since then, and is the source of the current angst among the "Five Eyes" countries (an intelligence alliance comprising Australia, Canada, New Zealand, the United Kingdom, and the United States), since companies and researchers in China now have to proactively look for military applications of their innovations.

On the other hand, Premier Li Keqiang has been calling for "mass entrepreneurship and mass innovation" on various occasions. In May 2015, under his leadership, the State Council unveiled "Made in China 2025," a policy document aimed to comprehensively upgrade Chinese manufacturing (State Council of the PRC 2015). Influenced by (and somewhat resembling) Germany's "Industry 4.0" strategy, the document lays out a "three-step" strategy, having objectives of turning China into a manufacturing powerhouse by 2025; placing China in the medium ranks of global advanced manufacturing by 2035; and solidifying China's advancing manufacturing status and being at the forefront of global manufacturing by the time the PRC celebrates its centennial in 2049. Subsequently, the State Council also has formulated another strategy focusing on the development of artificial intelligence (AI). It appears that advanced manufacturing and AI are the areas that China bets on in the coming decades for its economic restructuring and high-tech development.

Box 3. Artificial Intelligence

China's advances in artificial intelligence (AI) were fully displayed during President Trump's visit to Beijing in November 2017. As President Trump addressed a tech conference in English, his voice was live-translated into Mandarin Chinese. iFlyTek, a Chinese AI company specializing in voice and speech recognition, translation, and surveillance, provided the translation software showcased at the conference. Established in 1999, iFlyTek produces 80 percent of all speech recognition technology in the country and, unsurprisingly, has strong ties with the Chinese government (Human Rights Watch 2017). iFlyTek's voice-based technologies have been used in hospitals to direct patients to appropriate departments based on their symptoms, and in cars to allow drivers to communicate with their smartphones through voice commands (Sun 2017). With over 500 million users in China, iFlyTek is able to gather a significant amount of data through everyday interactions with its users (Sun 2017).

China's loosely enforced privacy laws and strong governmental support enable AI companies like iFlyTek to gather and access data at a much faster pace than its international competitors. Chinese citizens are also less averse to having their data taken by the government than their Western counterparts. All of this in turn gives Chinese companies a competitive edge in developing its AI technologies. There remain, however, serious privacy and security concerns regarding the use of AI. A 2017 report by Human Rights Watch indicated that iFlyTek is collaborating with Chinese authorities to pilot a surveillance system using voice recognition to identify individuals of interest from phone conversations (Human Rights Watch 2017; Mozur and Bradsher 2017).

At a time when the US administration is cutting funding to the National Science Foundation, China is investing heavily with the Ministry of Science to put tens of billions of yuan into supporting a new AI project (Wang 2017a). In addition to government funding, China's tech giants have also jumped into the fray. Baidu has opened two AI research labs in Silicon Valley (Baker 2017); and Tencent has opened an AI research lab in Seattle in addition to its AI lab in Shenzhen (Jiang 2017; Soper 2017).

PRIORITIZATION AND RESOURCE CONCENTRATION

Prioritization to achieve quality and enhance performance is nothing new; they are especially important for China given its resource constraints. In the mid-1990s, China rolled out a principle of "anchoring one end, freeing up the other," attempting to redefine the role of the state in the support of research through resource allocation and concentration. This involved a radical reduction in the traditional role of supporting applied R&D in industry and a more concentrated focus on areas of science most susceptible to "market failures" – basic research, education, pre-competitive R&D, and research for public missions. The reform of State Council-affiliated research institutes starting from the late 1990s corresponded to this principle. Then-President Jiang Zemin also put his weight behind the emphasis with his remark on "doing what we need and attempting nothing where we do not."[2] This means that China had to be selective in supporting various research endeavors and research personnel to best utilize rising but still scarce resources. In retrospect, the initiation of the national S&T programs such as the State High-Tech R&D (863) Program in 1986 and the State Basic Research and Development (973) Program in 1997, among others, reflects such strategic thinking.

In the early twenty-first century, the formulation of the 15-year MLP was essentially an exercise to identify priority areas and programs. As mentioned in the previous chapter, among its 11 broad "key areas" pertaining to national needs and eight "frontier technology areas," the MLP further identified a series of priority projects. For instance, under the "key area" of new materials, the plan included work on smart materials, high-temperature superconducting technology, and energy-efficient materials. In addition to the priority areas, the MLP further identified a series of large national "mega-programs" in engineering and science. The inclusion of these programs was one of the more controversial aspects of the plan, although it did continue the legacy of science planning, especially the lasting influence of the strategic weapons programs (or *liangdan yixing* model), on Chinese thinking (Cao et al. 2006; 2009).

The "Made in China 2025" policy is perceived as a departure from the indigenous innovation rhetoric displayed in the MLP as well as the Strategic Emerging Industries (SEI), another initiative prioritizing seven critical high-tech industries (US–China Business Council 2013),[3] both under Hu Jintao and Wen Jiabao, Xi Jinping and Li Keqiang's predecessors. However, commonality and continuity with the previous strategies/initiatives/plans also are significant. That is, the policy still highlights similar priority sectors, from new advanced information technology, automated machine tools and robotics, new materials, to biopharma and advanced medical products. It also calls for utilizing indigenous or China-based patents, inventions, and ideas, rather than those imported from abroad and skillfully adapted (Kennedy 2015). For example, it is expected that China will increase the use of domestic components in production to 40 percent by the year 2020 and as high as 70 percent by 2025. In other words, the role of the state in planning and prioritization has been both apparent and ubiquitous.

CHINA'S STATE-LED HIGH-TECH DEVELOPMENT AND INNOVATION: SOME CHALLENGES

In its burgeoning period, China's high-tech development was mostly chaotic but bottom-up and spontaneous. Back in the early 1980s, seeing the high-tech proliferation in Silicon Valley and along Route 128 in Massachusetts in the US, some of the earliest returnees wanted to explore means of technology diffusion in China. Universities and R&D institutes started to spin off enterprises with their research achievements, especially in information technology, in Beijing's Zhongguancun, where there is a concentration of talented scientists and students (see, e.g., Lu 2000; Segal 2003). Such a bottom-up effort combined with the political leadership's modernization drive, thus also leading to a reform of the S&T system that could better serve the nation's economic development. The first science park started to take shape in Zhongguancun, which, as noted previously, gradually became known as China's "Silicon Valley" (Cao 2004). Now, high-tech development seems to be more top-down – orderly but caring more about mature enterprises in Zhongguancun as well as in China's other 167 high-tech parks, where it is easier to find a first-rate infrastructure than to nurture a first-rate high-tech startup, let alone genuine entrepreneurial spirit. China's innovation, and indeed, its economic growth, has been state-led (Unger and Chan 1995; Huang 2008).[4]

Such a state-led model of technological development has its merits in concentrating and mobilizing resources to fulfill certain goals. For example, it did help China to develop its strategic weapons programs in the 1950s onward. So, time and again China has followed in the footsteps of its past success, and the MLP was formulated exactly with this model in mind (Cao et al. 2006; 2009). However, there are some inevitable problems associated with China's state-led innovation model.

First, much of the government's intervention in China is not about basic science and research on public goods but about innovation, mostly in the domain of enterprises. To put the Chinese case in perspective, debates have gone on in many countries for some time about the utility of national, state-led programs of innovation, in contrast to the belief that decentralized, market-responsive approaches are far more successful (Cao et al. 2009). Governments invest in R&D because of the "public good" nature and out of concerns about market failure (Mowery 2009; Stephan 2011; Wong 2011). With its long-run horizon for returns and uncertainties, R&D, especially basic research, receives less attention and investment from enterprises, thus making intervention from the state necessary (Nelson 1959; Arrow 1962). The "entrepreneurial state" also tends to take risks by creating a highly networked R&D system for the national good over a medium- to long-term time horizon (Mazzucato 2013). Indeed, there is a need to recognize a proper role for the state in promoting new knowledge and techniques, but determining what is "proper" remains contentious, and varies from country to country (Cao et al. 2009).

Second, the Chinese government is involved in picking winners, prioritizing industries, and betting on manufacturing stars.[5] High-tech firms have to be certified by the government, which in turn grants preferable policies toward them. As a result, some of the so-called high-tech enterprises are just "high-tech" in name – rather, they are government's pet priorities. Instead of using invisible hands to promote innovation, governmental efforts have been project-oriented and have supported and evaluated enterprises according to economic but not necessarily innovative indicators.[6] Consequently, the majority of enterprises do not choose to be innovative for the sake of their own development, but rather to respond to the government's policy incentives. Enterprises that receive funding as subsidies from government may not have the pressure to innovate, since innovation may not be their spontaneous and independent choice. Therefore, government funding often results in wasted money rather than significantly innovative outcomes, and vicious cycles of bubble and burst.

Third, with local governments fiercely grabbing land for developing various "parks," including high- and new-tech ones, it is not uncommon that these "parks" see concentrations of manufacturing products that are high-tech in name but low-end and low-tech in reality. As the end-users of such products are not necessarily located in China but in developed countries, any changes in the policies in terms of tax credits and subsidies elsewhere have influenced the production far away in China with widespread overcapacity just being one of the consequences. The development of the solar photovoltaic (PV) industry in China is one such example. Because of higher oil prices, there was worldwide expectation building around solar and wind energy, which led to what then-Premier Wen Jiabao referred to as the "blind expansion" at the 2012 annual session of the National People's Congress. But in the aftermath of the global financial crisis, Chinese solar PV companies saw their stock value plunging – some even went bankrupt – as a result of the overcapacity and change of policies toward the industry elsewhere. The new concerns are whether robotics, big data, cloud computing, among others, which the state has anchored in "Made in China 2025" and other policy initiatives, could become a new battlefield for the competition for resources and government support, although it is predicted that this time is different as there has also been much private investment. For instance, at present, there are already several hundred various kinds of "parks" focusing on new energy, LED, advanced manufacturing, and dozens of high-tech parks for big data and cloud computing, which again unfortunately are mainly engaged in assembly rather than manufacturing based on advanced and sophisticated innovation. The government's policy guidance for innovation – indeed, bureaucratic intervention – may well prevail over enterprises' awareness of the best practice in which to innovate. Given the state's performance evaluation orientation, high-tech parks still mostly embrace FDI since it will generate GDP and employment far more rapidly than Chinese firms' entrepreneurial efforts at indigenous innovation.

Fourth, the state has yet to formulate coherent policy measures specifically geared toward the development of high-tech parks and enterprises within parks; it therefore lacks the power to consistently and effectively intervene in such efforts. For example, in the early stage of designing the development strategy for national high- and new-tech zones, some scholars proposed schemes from complete laissez-faire to complete government control, with selective support to big cities such as Beijing and Shanghai and others in between. They favored small-scale experiments in a few Chinese cities to first test the Silicon Valley model (framed, of course, as having "Chinese characteristics"). But before a decision had been made as to what scheme to introduce, high-tech parks had already been built throughout the country, making it virtually impossible to reverse the trend.[7] Many of the national high-tech parks have been used as vehicles to attract FDI and projects on the state's priority list, in order to meet the need of the local economy rather than clustering truly innovative startups. Policies and approaches that have served other economies well have proved unsuitable for China – and, importantly, policies that served China well during its catch-up period may not necessarily do the same in its efforts to leapfrog to the forefront of innovation.[8] Once policies are made, they tend to become embedded; policies that no longer are applicable are seldom officially scrapped, thus furthering confusion and uncertainty. It is probably from this perspective that Xu Guanhua, an avid champion of the state-led innovation when he was China's minister of science and technology, lamented that the premise of the model is wrong and enterprises with the mentality of responding to government promotion policy are likely to be just mediocre (JRJ 2012). Other S&T administrators have had similar strikingly different views toward China's S&T and innovation policy, both when they were on the job and after their retirement.

In sum, although innovation is now officially a top priority and a key driver of China's next stage of economic development, at the same time the government's top-down approach puts innovation into an institutionally uncertain environment, where innovators have to skill-fully and tactically chart their development trajectory, mindful of any possible policy change all the time (Breznitz and Murphree 2011). Such an environment also tends to favor state-owned enterprises (SOEs), mature enterprises, and foreign-invested enterprises, while private small and medium-sized enterprises (SMEs) and startups are facing chal-lenges not only in attracting talent and capital, among others, but also in an unpredictable policy environment. The state actively channels public resources to embark on strategic development initiatives, while retaining controlling stakes in the enterprises that dominate strategic and resource sectors. This has been at the expense of the largely market- and private-sector-driven development in the coastal provinces, resulting in a strong disincentive for SMEs to innovate.[9] As a result, there is a weak environment in support of innovation at the enterprise level. The recent development of "SOEs advancing and non-SOEs retreating" is just one example of the dilemma confronting China's high-tech development (Xu 2015).

GOVERNANCE AND ACCOUNTABILITY

The significant expansion of government funding for research and innovation, as seen in recent decades, is raising new concerns about the performance of the research system and whether national resources are being used wisely. While China still is a long way from democratic accountability, there is no doubt that criticisms of the system, and frequent reports of fraud and other types of misconduct in the technical community, including the recent scandal involving the "China chip,"[10] are raising questions in the NPC, the Ministry of Finance, and elsewhere about the public administration of science in China and the management problems of government agencies. To their credit, the Ministry of Science and Technology and the National Natural Science Foundation of China have responded quickly to recent cases of misconduct and fraud. The new reform measures on project evaluation and budgeting procedures have been introduced to monitor research more closely and prevent and punish fraud and other forms of deviant behavior in scientific research. But it remains to be seen whether these procedures are enforced to safeguard academic integrity. For instance, governance problems have arisen with the emphasis increasingly placed on technology transfer, as cases have been settled or pending involving scientists who channel public R&D funds to their own companies or companies with which they are associated. Chinese scientists also need to be accountable for the amount of public funds for research by producing significant breakthroughs rather than merely minor outcomes.

CONCLUSION[11]

Despite the ideological legacy of socialism and central planning, after some four decades of reform, the political economy of science and technology has become much more complex, especially with the important roles that MNCs play in research and innovation in China, and because of the growth of new technologically progressive Chinese firms, many of which have no, or only tenuous, relations with the state. On the one hand, the roles played by these two dynamic sectors suggest that the course of Chinese technology and innovation development over the coming years cannot accurately be defined, simply, as state-led innovation. On the other hand, China's policy framework and processes clearly privilege the institutions of the state and make it more difficult for MNCs and new technology enterprises, also stakeholders in the national innovation system, to be represented. Thus, the persistence for state-led innovation may run the risk of marginalizing two of the more technologically dynamic sectors in society. But the question does require that mechanisms for assessing and evaluating the costs and benefits of state action be employed. Market forces in China are beginning to help in these tasks, as is the increasing commitment to formal research evaluation, and the growing concerns expressed about government accountability also show promise.

Efforts to promote state-led innovation also must face underlying cultural factors. Again, the state is aware that much needs to be done to generate the growth of a "culture of innovation" in China, but it remains unclear whether this can be accomplished by state action. The cultural problem has been linked to the persistence of Confucian values, stressing obedience to hierarchy and harmony, and the ways in which such values come to shape the educational experiences of Chinese youth (see chapter 6). It also has been linked to state policies that encourage technological entrepreneurs to spend more of their time and energy cultivating good relations with officials rather than developing and marketing innovative products. Finally, pressure from the state through investment and rhetoric also may explain why there have been rising incidents of fraud and misconduct in science, or the goals of intimately linking scientific research to national pride by catching up with and surpassing currently leading countries may sometimes lead to unintended and mostly undesirable consequences (Segal 2011).

In terms of sheer numbers, China's overall performance in terms of publications and patents is impressive. Measured by the number of papers published in all journals indexed by Science Citation Index (SCI),[12] in 2016 China ranked second in the world, accounting for some 20 percent of the world's total (National Natural Science Foundation of China 2017). Chinese inventors also have filed the most patent applications in recent years (Fang et al. 2016). China's State Intellectual Property Office (SIPO) recently released its 2017 report on patent activities in China. According to the report, the number of patent applications increased 14.2 percent from 2016, registering 1.38 million, while the number of patents granted reached 420,000, with a strong concentration in China's east coast region (P. Li 2018).[13] According to *Science & Technology Indicators*, recently released by the US National Science Foundation and National Science Board, "China has become − or is on the verge of becoming − a scientific and technical superpower," in terms of R&D spending, technical papers, and its technical workforce (Samuelson 2018).

It is unclear whether quantitative measures in fact represent an upward trend in the quality of Chinese science and innovation, or whether money is to some extent wasted by simply producing high numbers of papers and patents. As we argue elsewhere (see chapter

6), Chinese institutions of higher learning still focus on quantity rather than quality as key performance indicators. While quantitative measures provide objective assessment criteria that reduce the chance for performance evaluation to be manipulated or involve impersonal factors, they also invite an overemphasis on quantity rather than quality. The lack of full, democratic accountability may deprive China of an important source of information and feedback as to whether that issue is being tackled (Cao and Suttmeier 2017).

<table>
<tr><td>4</td><td>

China's International S&T Relations
From Self-Reliance to Active Global Engagement

</td></tr>
</table>

Only if core technologies are in our own hands can we truly hold the initiative in competition and development. Only then can we fundamentally ensure our national economic security, defense security and other aspects of security... On the traditional competition field of international development, the rules were set by other people... To seize the great opportunities in the new scientific-technological revolution and industrial transformation, we must enter early on while the new competition field is being built, and even dominate some of the construction of the competition field, so we become a major designer of the new rules of competition and a leader on the new field.

PRC President Xi Jinping Speech to the Chinese Academy of Sciences/Chinese Academy of Engineering, Beijing, June 2015

INTRODUCTION

The Belt and Road Initiative (BRI)[1] Summit held in Beijing in 2017 has proven to be a golden opportunity for China to showcase its willingness and capability to fully participate in global governance and S&T affairs. It also seems to have served as a transformational moment in terms of China's larger overall global role. The BRI proposes to provide physical connectivity, promote free trade, and enable policy coordination among the more than 60 countries that will be involved. The Chinese government has been particularly proactive in trying to establish international S&T and innovation cooperation networks under the initiative's overall framework. Bai Chunli, current President of the Chinese Academy of Sciences, told a State Council press conference that CAS was determined to promote the creation of the BRI S&T cooperation network by 2030, and that the network will play a dominant role in China's future bilateral, multilateral, and regional S&T cooperation (Bai 2017a). The joint report on BRI S&T cooperation, issued by the National Science Library of CAS and Elsevier, clearly states that China sits at the hub of the network in terms of scale, capability, and input of R&D activities, enabling the PRC to effectively use its many advantages to facilitate wider cooperation among participating countries with vastly different S&T development levels (CAS 2017). Allowing for such differences, the initiative already has resulted in joint laboratories, collaborative programs, cooperative R&D, personnel exchanges, and, most importantly, Chinese technology assistance.

If the BRI is a type of promissory note for future international S&T collaboration, China's recent performance track record suggests that it will likely deliver. As of 2015, the Ministry of Science and Technology (MOST) had funded over 12,000 cooperative research programs, involving more than 20,000 participating personnel; China's National Natural Science Foundation also has funded cooperative research programs (see figures 4.1 and 4.2). Another indicator is the number of co-authored SCI papers. By 2014, fully a quarter of China's SCI papers had international collaborators, two-thirds of which listed the Chinese contributor as first author. This strongly suggests that Chinese scientists are playing an increasingly central role in internationally cooperative research projects, particularly in the fundamental research fields where such papers are concentrated (People.cn 2015).

China's engagement in international S&T affairs began with the founding of the PRC in 1949, when the CCP formulated and implemented a bilateral S&T cooperation agreement with the former Soviet Union (*yi bian dao*) – a relatively short-lived arrangement that was followed by the policy of self-reliance (*zi li geng sheng*) in response to

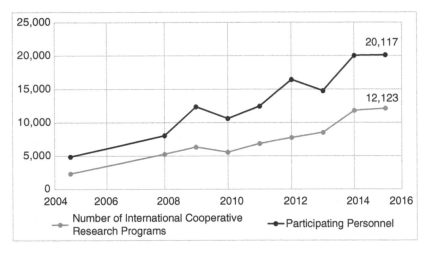

Figure 4.1 International S&T Cooperation Programs Funded by MOST
Source: Compiled with data from MOST. *China Statistical Yearbook on Science and Technology 2016*. Beijing: China Statistics Press

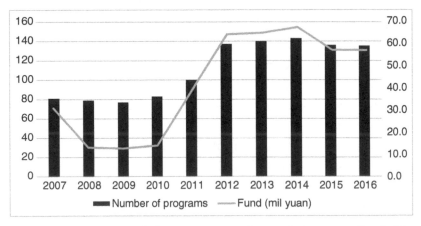

Figure 4.2 Overseas Scholar Cooperation Research Programs Funded by NSFC (funds are in million yuan)
Source: Compiled with data from *Annual Report of the China National Natural Science Foundation*, Beijing, 2007–16

Moscow's termination of technology assistance in 1960. The relationship between Moscow and Beijing had been highly asymmetrical as China was very dependent on the former USSR for massive inflows of industrial equipment and managerial know-how to jump-start the Chinese economy after the end of the civil war with the Kuomintang in 1949. Beginning in the late 1970s, following the turmoil of the Cultural Revolution and beginning with Deng Xiaoping's Reform and Open Door policy, China's leadership shifted its focus to economic development. In terms of China's international S&T relations, guidelines were adopted to lay the foundations for managing China's foreign affairs, including international S&T cooperation. By the turn of the twenty-first century, China had achieved full-scale implementation of an international S&T cooperation system focused on acquiring foreign technology and fostering cooperative arrangements with international scientific institutions. With the Open-Door policy and the abandonment of the policy of self-reliance, China joined numerous international S&T organizations, and promoted foreign plant, equipment, and technology imports. During the first decade of the twenty-first century, the government has pushed for mutually beneficial international S&T cooperation, developing well-articulated programs in an effort to achieve this objective. China has played an increasingly active role in international organizations encompassing major science and engineering programs, while at the same time strengthening technical assistance to developing countries (Cheng 2008; Xu 2008). Since 2012, China has sought to plan and promote innovation with what it now characterizes as a global vision, embodied in various key national policies.[2]

China is now in the process of transforming itself from a technology importer to a technology exporter, as it pursues the indigenous innovation strategy discussed in previous chapters. In this chapter we focus on the successes and challenges in China's pursuit of its global S&T relations, which are central to its efforts to move from imitator to innovator. By 2017, China had established S&T cooperation partnerships with 155 countries and regions and executed 107 inter-governmental agreements on S&T cooperation. In addition, the PRC has joined over 200 inter-governmental international S&T cooperation and research organizations. It has appointed 144 S&T diplomats for its 70 overseas offices in 47 countries. And, as of the beginning of 2014, close to 400 Chinese scientists are currently holding office in international and S&T-related NGOs, including 28 as Chairperson and 50 as Vice-Chairperson. Among the world's 48 major cross-border big science programs and projects, four have been initiated by China and 17 have official Chinese participation; China also serves as an observer in three programs. This all demonstrates that China's presence in the structure and organization of global S&T governance is meaningful and steadily increasing.

THE ADMINISTRATIVE STRUCTURE OF CHINA'S INTERNATIONAL S&T POLICIES AND ENGAGEMENT

The S&T governance structure for China's international S&T engage-ment is composed of a number of key agencies and organizations. There are multiple ministries, commissions, central, and local gov-ernment entities involved in this sphere of activity. In this section, we focus on three that have emerged as the most important in organiz-ing and managing China's international S&T relations: the Ministry of Science and Technology (MOST), the Chinese Academy of Sci-ences (CAS), and the China Association for Science and Technology (CAST).[3]

Ministry of Science and Technology (MOST)

The Ministry of Science and Technology is the predominant entity that plans and implements China's overseas S&T affairs, providing the overarching framework for international S&T cooperation at different levels and by increasingly diverse actors. Since its mission is to foster economic growth and technological advance, MOST coordinates basic research, frontier technology research, and the development of key and advanced technologies. It also is mandated to formulate policies on international S&T cooperation and exchanges through bilateral and multilateral channels, guiding relevant departments and local governments in international interactions, appointing and supervising S&T diplomats, and facilitating assistance to and from China. MOST's Executive Office is responsible for drafting and formulating important policies, and handling tasks assigned by the State Leading Group for Science, Technology, and Education. A number of other departments play key roles in China's S&T development, commercialization, and foreign relations:

◆ The Department of Policies, Regulation and Supervision conducts research on S&T affairs, and in general provides guidelines intended to create an environment whereby S&T activities facilitate technology transfer − the transformation of research findings into commercial winners. In keeping with this mission, the department also organizes research on measures to reform the current S&T management system, drafts laws and regulations which are intended to strengthen intellectual property right protections, and examines the establishment and performance of research institutes.

◆ The Department of Innovation and Development is responsible for integrating and drafting plans for S&T development and innovation capacity building, evaluating innovation performance of research institutes, promoting regional S&T development, and guiding local S&T affairs.

◆ The Department of Resource Allocation and Management puts forward advice on allocation, optimization, and integration of S&T resources, drafts policies on key S&T project investments and R&D fund management, examines and verifies new S&T plans and funding at national level, and prepares budgets.

+ The Department of Basic Research manages R&D resource sharing across relevant departments. To accomplish this, it provides advice on official research policies and priorities, coordinates planning, participates in formulating key national S&T projects, selects project management agencies for key R&D programs, and engages in forecasting the needs for basic research.

+ The Department of International Cooperation is without question the most important department that bears the responsibility for China's international S&T engagement. It drafts policies on international S&T cooperation and exchange, providing guidance for the international S&T affairs of relevant agencies and local governments. For example, the department organizes inter-governmental innovation dialogues; bilateral, multilateral S&T cooperation agreements and exchanges; tracks country-specific deployment of key S&T programs; conducts technology forecasts; and promotes the construction of international S&T cooperation bases.

+ In March 2018, the State Administration for Foreign Expert Affairs (SAFEA) was placed under the oversight of MOST. For the last several decades SAFEA has been responsible for bringing to China a broad range of experienced scientific and technical experts from abroad to assist their Chinese counterparts with various developmental problems and issues. It also has sent large numbers of PRC delegations abroad, especially to the US, Western Europe, and Japan for training in management and an assortment of technical fields.

Chinese Academy of Sciences (CAS)

The Chinese Academy of Sciences, currently directed by President Bai Chunli, is structured as a comprehensive, integrated R&D network. It is the nation's high-end think tank, a merit-based learned society as well as a system of higher education, and has long functioned as the linchpin of China's national and global S&T ambitions. By 2016, CAS has 12 sub-academies, three universities, and more than 210 field observation stations, 130 national laboratories and engineering centers, and 100 research institutes. It also oversees over 20 national S&T infrastructure projects. The CAS is constantly undergoing reform and change, with mergers and consolidation of institutes becoming more and more common.

Since its inception, the CAS has made significant progress in fostering international S&T cooperation relationships. It has succeeded in developing extensive and diverse partnerships with research institutes and scientists across the globe, and is well positioned to play a central role in shaping China's S&T diplomacy from a substantive point of view. To take some recent examples (Bai 2017b), CAS has

- set up 20 collaborative groups with the German Max Planck Society in areas including astronomy, life sciences, and materials science;
- implemented several talent programs (such as the CAS Fellowship for Senior and Young International Scientists), attracting over 1,000 foreign scientists and engineers to conduct R&D activities at its institutes;
- initiated a BRI action plan in 2016 calling for international S&T cooperation, training and cultivating more than 1,800 S&T management and high-tech personnel for relevant countries; and
- plans to become the spearhead and central hub for a Asia-Pacific, Eurasia, and Asia-Africa collaborative innovation network system.

The structure and organization of CAS is well developed, with a number of departments responsible for managing domestic R&D programs and international S&T cooperation.

- The Bureau of International Cooperation's responsibilities include formulating strategies, plans, rules and regulations for CAS international cooperation and exchanges; coordinating academy-level international cooperation affairs; initiating and managing key cooperative programs and fellowships; and maintaining links with related agencies dispatched by international organizations in China.

✦ The Bureau of Science and Technology for Development is responsible for CAS's intellectual property rights (IPR) management; promoting IPR applications and R&D transformation; coordinating S&T personnel exchanges with local governments and the corporate sector; and conducting research on S&T-driven development strategies and international cooperation plans.

✦ The Bureau of Major R&D Programs is responsible for developing and managing key national S&T priority programs; implementing and coordinating CAS strategic and high-technology innovation initiatives; and construction of key labs and platforms.

China Association for Science and Technology (CAST)

Founded in 1958, the China Association for Science and Technology is under the direct jurisdiction of the Secretariat of the CCP's Central Committee. Its role includes promoting S&T exchanges and indigenous innovation, protecting and advancing the interests of science workers, organizing S&T professionals to participate in formulating national S&T policies, and facilitating non-governmental international S&T exchanges and cooperation through developing liaison with foreign S&T associations and scientists.

CAST is made up of national scientific and professional societies and local S&T associations. Among the national societies, 42 are in the natural sciences, 73 in engineering, 15 in agriculture, 26 in medical sciences, and 23 in interdisciplinary scientific fields. Local associations – totaling around 3,000 – include those organized by provinces, autonomous regions, and municipalities directly under the central government, cities, and counties. Among its various departments, the Bureau of International Liaison is mainly responsible for international S&T affairs. It is responsible for working out annual plans and advice for

CAST bilateral communications, conducting research and summarizing experiences on S&T exchanges, and exploring and developing partnerships with S&T associations in key countries and regions.

CHINA'S INTERNATIONAL S&T POLICIES: CONTINUITY AND CHANGE

Since Deng Xiaoping's Opening Up and Reform, the Chinese government has been consistent in encouraging Chinese companies to go abroad to better leverage international S&T resources, formulating a series of policies to guide its S&T engagement with other countries. These policies reflect the emphasis, discussed throughout this book, on strengthening indigenous innovation. From China's standpoint, indigenous innovation is necessarily coupled with an outward-looking strategy that calls for S&T partnerships and international collaborations. International S&T relations are thus best understood as constructed to serve China's goal of becoming a global innovation competitor, if not a leader. China's state-led efforts to achieve indigenous innovation have not been well received by China's rivals. The MLP, for example, was roundly denounced in a US Chamber of Commerce-sponsored report bearing the title, *China's Drive for Indigenous Innovation: A Web of Industrial Policies* (McGregor 2010). The report accused China of "hunkering behind the 'techno-nationalism' moat," switching "from defense to offense" in light of its economic ascendance as well as its fear of foreign domination. The MLP, according to the report, "is considered by many international technology companies to be a blueprint for technology theft on a scale the world has never seen before." The report obviously contains a great deal of hyperbole; nonetheless, the MLP's policies did provoke a strong reaction from China's major trade and technology partners. Given that innovation capability and talent increasingly drive competition among countries, China's leaders recognize that strong domestic S&T capacity has become the core requirement for meaningful and productive bilateral and multilateral S&T cooperation. For China, the emphasis on indigenous innovation has no longer meant self-reliance as was the case in the 1960s. Rather, it always has been seen as a pathway to strengthen China's leverage in the international technology market.

Budgetary allocations for international S&T cooperation have grown apace with domestic S&T spending, especially at the local level. As suggested above, China's emphasis on indigenous innovation should not obscure the fact that the government has spared no efforts to deepen and enlarge bilateral and multilateral S&T partnerships. The 13th Five-Year Plan,[4] in contrast to its predecessors, designates tasks and goals that serve Beijing's current strategy of science diplomacy, transforming itself from passive recipient to active donor as a central part of China's foreign policy (see figure 4.3).

China's international S&T cooperation strategy is carefully differentiated according to a categorization of partners into developed, developing, and neighboring countries. The Plan calls for increased openness

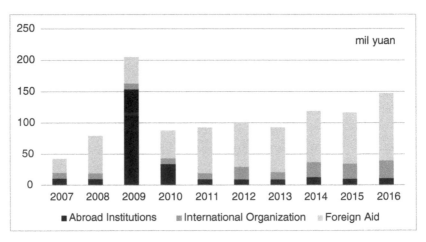

Figure 4.3 Distribution of Annual Expenditures by MOST
Source: Compiled with data from MOST annual expenditure reports

of China's national S&T programs, including offering governmental support to overseas experts who are expected to take the lead – or at least participate in – national S&T program strategic research. It also calls for deepening international cooperation on an equal basis with international partners (a claim which has been met with some skepticism). To achieve its goals, China has initiated and organized significant international science programs and projects; has become more actively involved in helping to set global S&T agendas; has accelerated the sharing of global large-scale scientific research information; and begun active participation in global S&T governance, including the formulation of international S&T cooperation rules. Chinese scientists have increased their participation in scientific exchange programs, as well as sought official positions in major international scientific and technological organizations. China's most recent – and clearly most dramatic – diplomatic move in the science field is the BRI S&T cooperation network, which calls for promoting technology transfer and assisting countries in training young scientists – a clear indication that China plans to play a central role in the international S&T landscape as a technology exporter as well as importer.

In January 2018, President Xi presided over a meeting of the Leading Group for comprehensively deepening reform of the central government. One resolution called for actively initiating and organizing international big science programs and projects;[5] another for strengthening regulations in IPR transfer. China is clearly looking outward, as it plans its S&T future.

PREPARING CHINA'S HIGH-END TALENT: OVERSEAS STUDY

Any examination of China's international S&T relations inevitably must address the issue of overseas study by Chinese students and scholars. The rapid growth in the numbers of PRC students going abroad serves as an indicator of the potential impact that returned students and scholars have had and will have in the future development of S&T in China. In 2006, 134,000 Chinese students went abroad; by 2016 that number grew to over 544,500. Each year the number has been steadily expanding (see figure 4.4). While during the 1980s and 1990s, the bulk of students focused on science and engineering fields, an increasing percentage began to study law, business, finance, and economics after 2000. And, more recently, there has been a growth in the number of students interested in studying social sciences and humanities, thus giving a more rounded picture to the development of Chinese talent in overseas institutions.

The most critical issues surrounding these students is whether or not they return to China and how they are utilized once they return home. We can see some major changes in the nature of the groups going abroad. In the 1980s and early 1990s, the bulk of students going abroad were *gongfei*, or government sponsored, but today the majority

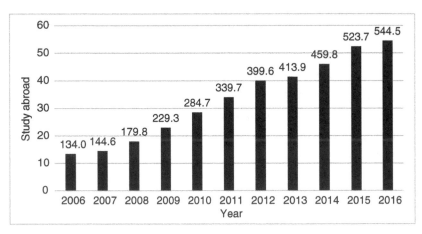

Figure 4.4 Number of PRC Students Going Abroad for Study (in thousands of students)

Source: Compiled with data from annual reports by Ministry of Education (PRC) on Chinese students who study abroad and return

of students are *zifei* or self-funded. In addition, during the initial period, the majority of students seeking overseas education were graduate students pursuing masters and PhD degrees on domestic and foreign scholarships. Today, more are undergraduates going abroad for study and paying their own way. Initially, there was a fairly large brain drain as many students preferred to remain overseas to build their careers. This trend was reinforced after the June 4th incident in 1989, when many students feared returning home because of the political environment in China. That situation gradually improved and a larger percentage of students decided to return after graduation or some limited work experience abroad. Things changed decisively after 2008 and the global financial crisis, as it proved to be more and more difficult for professionals to gain permanent resident status in countries such as the US due to the economic situation (see figures 4.5 and 4.6). Under the Trump administration, the situation has become even more difficult in the US regarding H1B or work visas; therefore, it is likely more and more PRC students will return home after graduation. Of

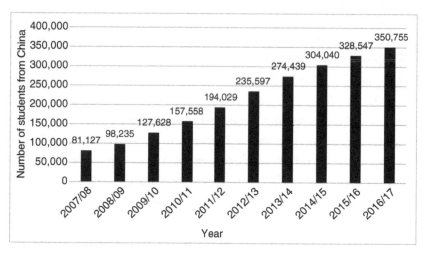

Figure 4.5 Number of PRC Students Going to the USA for Study
Source: Compiled with data from IIE Open Data 2017

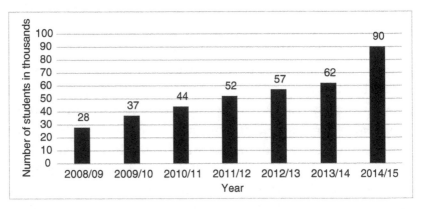

Figure 4.6 Number of PRC Students Going to the UK for Study (in thousands)
Source: Compiled with data from Statista

course, there also are strong pull factors coming from the PRC side of the Pacific Ocean, including the rising economy, the increased opportunities for entrepreneurial ventures, and the rise of the so-called shared economy.

The long-term reverse of the so-called brain drain could have significant implications not only for the development of China's high-end talent pool, but also for helping to engender the types of reforms that are badly needed to improve the work environment for R&D talent and to spark changes that would help support the rise of a new class of technological entrepreneurs. These returnees, some of whom already have become catalysts for change inside Chinese companies, could be important instruments for bringing new ideas and concepts into research and corporate settings across China. It would be ideal, of course, if these returnees would return to a broad array of locations across all of the Chinese economy, but so far there has been a tendency for returnees to be attracted primarily to key coastal cities such as Beijing, Shanghai, and Shenzhen.

The more immediate question, however, is whether or not these returnees become disillusioned with the environment in which they find themselves. Up to now, the experiences of returnees have been mixed, with many simply finding too many points of resistance within the Chinese R&D operating environment when compared to their work experiences in the US or the UK. While talent-attracting policies enacted under both Hu Jintao and Xi Jinping have been directed at overcoming problems caused by nepotism, personal relationships (*guanxi*), seniority bias, and gender bias, these problems remain real and ever-present. Also, it is safe to say that a real meritocracy has yet to emerge fully inside of the scientific establishment and the corporate world. This is a real deal-breaker for many Chinese considering a return to China and building their career inside a Chinese research or academic institution. Progress is certainly being made, but sometimes at a snail's pace the further one gets away from coastal China. Installation of a true performance, merit-based culture will determine to what extent and how fast the number of truly high-end returnees will increase (Li 2005; Zweig 2006; Simon and Cao 2009a; Wang et al. 2011; Zweig and Wang 2013).

FOREIGN INVESTMENT AND TECHNOLOGY ACQUISITION

It is important to recognize that enterprise-to-enterprise cooperation in the form of foreign investment has been seen by China as a major vehicle for acquiring technology. The decision to initiate a new joint venture law as part of the Open-Door policy was made not only for job creation and to earn foreign currency, but also to assist in the transfer of technology from the world's leading multinational firms to China. Of course, for many foreign firms, their main goal in coming to China initially was to harness cheap Chinese labor and later to address the rapidly expanding domestic Chinese market. In this way, the basic equation for business-to-business cooperation was established – market access in return for access to technology. Unfortunately, while foreign investment did bring jobs to China, especially along the 14 cities on the PRC's east coast to start with, and it did generate badly needed foreign exchange, it did not bring the large-scale visible technology transfer the Chinese government had hoped for.

The reality of China's foreign investment experience has been that the transfer of hardware has been more prominent than the transfer of software, or high-end technical know-how. Within many joint venture projects, conflicts between the Chinese and foreign partners often precluded possibilities for meaningful technology cooperation, especially given the weak IPR protection environment that existed in China for much of the 1980s and 1990s. What did get transferred – and should not be underestimated in terms of its long-term importance – was managerial know-how, quality control systems, and basic manufacturing knowledge. Japanese, Korean, and Taiwanese firms often have been criticized for not providing much in the way of advanced technology transfer through their foreign invested projects in China. Unfortunately, these criticisms ignore the large-scale transfer of manufacturing and managerial know-how that firms like Hitachi, Toshiba, Samsung, and LG brought to China over the years. The same thing can be said about US and European firms, many of whom also found themselves criticized for their weak offerings in technology transfer. Here again, however, while many projects conducted within Sino–foreign joint ventures were simply assembly or so-called "screwdriver operations," the large majority of projects did entail substantial training programs and managerial upgrading that contributed to making the so-called "factory to the world" model as successful as it was for the first 20–25 years of China's open-door era.

Most recently, within the context of the discussions about indigenous innovation, there has been some confusion about just how far foreign technology transfer has proceeded, especially regarding the data associated with the rapid increase in so-called "high technology" products from China. The PRC government's push on indigenous innovation stems largely from these data which indicate that despite the expansion of Chinese exports in information technology and electronics products, the bulk of the value added has not come from the Chinese contribution. In some cases, between 60 and 80 percent of the value of Chinese high-technology exports has come simply from the assembly and processing of parts and components imported from abroad. The most notorious case involves the Apple iPhone, which has been designed in Cupertino, California, but assembled largely in China, as the back of the iPhone shows. A 2011 study assessing the value added for an iPhone 4 and a 16GB Wi-Fi iPad, both introduced in 2010, found that Apple captured US$321 (80 percent) of the US$401 total value capture for the iPhone 4, and US$150 out of the US$238 (63 percent) total value capture for the iPad. China, on the other hand, was able to capture only US$10 (2.5 percent) for each iPhone and US$8 (3.4 percent) for each iPad (Kraemer et al. 2011; see also Sun and Grimes 2017).

This type of mismatch between Chinese technology priorities and the strategies of foreign firms has generated a great deal of anxiety inside the upper echelons of the Chinese leadership in Beijing. At the time, it also created the sense that in a world of increasingly rapid innovation, Chinese progress was actually lagging behind the West and that some type of strategic re-think was necessary to ensure a more equitable financial outcome from world-class products such as the iPhone that were being manufactured in China. What frightened Chinese leaders the most was one key fact: while countries such as the US and Japan were continuing to run a huge IPR trade surplus from profits generated by the licensing of their own technologies abroad, China was continuing to experience a large, ongoing IPR deficit – a US$10 billion deficit in 2009 according to a World Bank study (Ghafele and Gibert 2012) – because of the royalties and license fees paid for the dependence of Chinese firms on imported know-how. This had led to a variety of PRC government actions, often not welcomed in the West, that seek to close this gap by pushing hard on foreign firms to share their technology for access to China's now highly dynamic domestic market and its increasingly skilled labor force.

CHINA'S INTERNATIONAL S&T RELATIONS WITH MAJOR COUNTRIES

Under its government-to-government bilateral arrangements, numerous scientists and engineers have participated in a broad array of collaborative projects with their counterparts abroad. Starting in the 1990s, however, China has greatly expanded its international S&T engagements; more and more activities are now occurring outside the government bilateral accords and now include a rapidly expanding number of university-to-university ties, corporate linkages, and cooperation with think tanks. Most recently, China's provincial and local S&T organizations also have become increasingly involved in orchestrating overseas S&T ties; many Chinese provinces and municipalities are leading the charge to find new, dynamic international S&T cooperation partnerships.

Although China is extending S&T cooperation partnerships with an increasing number of countries globally, its focus is still on working with the major developed states, based on national recognition of the prevailing technology gaps.

China–US S&T relations

The 1979–89 period featured the inception of China–US S&T cooperation. The 1979 agreement on science and technology has functioned as the overall framework under which the two governments promote S&T cooperation in various forms and through a large number of channels. The two countries concluded an accord to allow for student and scholar exchanges as well. From 1978 to 1987, the number of students and visiting scholars sent by the Chinese government to the US reached 25,000 (Orleans 1988). The China–US S&T relationship is overseen by a Joint Commission that meets on a scheduled basis to review existing programs and identify new areas of cooperation. The membership on the Joint Commission reflects participation from the key government agencies tied respectively to China's State Council and the US Executive branch of government.

Bilateral S&T cooperation experienced rapid growth during the early years as it was new and exciting; the two parties invested significantly to support joint programs. By 1987, there had been 27 signed cooperative agreements. That said, China–US S&T cooperation during this period was also constrained by a variety of political and financial factors and was largely asymmetrical and one-sided because China concentrated on utilizing US-provided instruments and equipment, and experts from the US played the primary role in knowledge dissemination and personnel training. Nevertheless, it is important to bear in mind that the two sides also had quite different objectives. The US intended to counter the former USSR by developing rapport and trust with China, and the US technical community was interested in the distinctive natural and social phenomena in China. The Chinese, however, assumed that involvement with the US technology system (and with Western Europe and Japan) would be a useful mechanism for promoting badly needed modernization and closing the prevailing technology gap with the world's leading powers (Suttmeier 2014a).

From 1990 to 1999, bilateral S&T relations witnessed some apparent decline, followed by resumption of activity. Due to the events in Tiananmen Square on June 4, 1989, many programs were curtailed, including China–US space cooperation. The US also terminated high-level political exchanges and postponed meetings of the Joint Commission, which dealt a heavy blow to S&T cooperation. Gradual resumption of bilateral S&T cooperation began in 1994, when the two parties decided to restore the Joint Commission Meeting (JCM). With China's accession to the World Trade Organization and the smooth transition to the next generation of Chinese leaders – Jiang Zemin to Hu Jintao – the possibilities for a new growth spurt began to appear.

From 2000 to 2015, China–US relations were characterized by comprehensive and rapid development (Suttmeier 2014b). Then-President Hu Jintao remarked in 2012 at the 14th meeting of the Joint Commission that S&T cooperation had become an important driving force for Sino–US relations and a critical component of people-to-people exchange. This cooperation fell into six main areas: energy and physics, health and life science, ecology and environmental science, agriculture and food science, science education, and metrology. It is worth noting that beginning in 2006, when the MLP was launched, the agenda for bilateral S&T cooperation took on a heightened understanding of the critical need to more explicitly address interdisciplinary research issues, cutting-edge science, and pressing global issues, such as global climate change, clean energy, and carbon capture and aggregation. In other words, the rising salience of these global issues altered the context for both sides to think about how S&T cooperation might proceed. A series of new initiatives were taken that were based on high-level political communications (Springut et al. 2011). The Strategic Economic Dialogue (SED) that came into place in 2006 and its successor, the Strategic and Economic Dialogue (S&ED), produced an enormous expansion of activities and functions. The latter launched the Ten-Year Framework on Energy and Environment Cooperation in 2008, designating clean water, clean air, clean vehicles, and energy efficiency as key areas with high priority for cooperation. In 2013, the US State Department and China's National Development and Reform Commission (NDRC) co-chaired a joint climate change working group under the general framework of S&ED. By 2011, China had risen to become the top collaborating partner of the US, outpacing the UK, Japan, and Germany – nations that are long-time partners of the US in science (Suttmeier and Simon 2014). By the end of the decade, in jointly authored scientific papers, Chinese scientists claimed first author-ship much more frequently than US counterparts (Wagner et al. 2015).

One of the key elements of these new dialogues was the initiation of the China–US Innovation Dialogue, which actually began in 2008 as part of a discussion about how the Chinese side could improve per-formance of its own innovation. The Innovation Dialogue had great potential when it started because it might have served as a useful vehicle for exchanging meaningful information about the evolving requirements for successful innovation in the twenty-first century. Unfortunately, the Innovation Dialogue ended up being neither a real dialogue nor about innovation. On the US side, growing disenchantment with China in the US Congress led to constraints being placed on the White House Office of Science and Technology (OSTP) about expansion of S&T cooperation; funding was tightly controlled. Moreover, the innovation agenda was hijacked by the US Trade Representative Office (USTR) and made to focus on extracting concessions from the Chinese side on pressing trade matters. The bulk of discussions ended up concentrating on dismantling Chinese policies regarding the promotion of indigenous innovation. On the Chinese side, the prize still remained in sight, though their side also was often distracted from the innovation-related issues that they expected to drive the Innovation Dialogue.

While as of this writing it is too soon to tell how the Trump admin-istration will affect the overall China–US S&T relationship, certain things have become clear. First, with the weakening of the OSTP, including a long delayed appointment of a new director who usually holds a concurrent position of Presidential Science Adviser, the S&T relationship lacks a major policy advocate on the US side. Second, Congress remains reluctant to provide any substantial funding for growing the relationship in new areas. This is unfortunate because with China making real progress in terms of its S&T capabilities, there is now more opportunity than ever before to take advantage of the greater symmetry in the relationship. Third, because of tensions over trade, technology transfer, North Korea, and the South China Sea, the political environment does not support an expanded relationship, let alone maintaining the status quo. In fact, the bilateral S&T agreement has not formally been renewed as of this writing, as the two countries have come to loggerheads over some delicate issues regarding past and present IPR issues. In addition, the decision by the Trump administration in March 2018 to invoke special legislation under US Section 301 laws concerning trade and investment with China has threatened a US–China a trade war, with technology theft and other related IPR issues positioned at the center of American concerns. And finally, the reality is that non-government exchanges and cooperation regarding the private sector, universities, and think tanks have far surpassed the level of government-to-government cooperation. This was the main thrust, albeit implicit, of comments made by then Vice-Premier Liu Yandong, during her Fall 2017 visit to the US, where she highlighted the need for greater emphasis and support for people-to-people diplomacy in the area of China–US S&T cooperation. It also became the focal point of critical comments by the Director of the Federal Bureau of Investigation in early 2018, when he warned American higher education institutions about the vulnerability of their institutions to "non-traditional" collectors of critical scientific and technical information coming from China (Woodruff and Arciga 2018; Yam 2018).

China–Russia S&T relations

China–Russia S&T relations should be divided into three phases – Phase One: close China–Soviet S&T cooperation (1949–60); Phase Two: the Sino–Soviet Split (1961–90); and Phase Three: renewed China–Russia S&T cooperation in the post-Cold War era (1991–present). Relations today between the two countries under Russia's Vladimir Putin and China's Xi Jinping respectively seem to be on the verge of a golden era, as they both see expanded opportunities for building the bilateral S&T relationship.

During Phase One, the former Soviet Union transferred technologies to China that helped lay the foundation for the renewal of industrial production, assisted China with formulating a 12-year plan for S&T development, established S&T research and design institutes, developed scientific research and industrial technology, and cultivated a group of S&T talents (Jersild 2014). That said, the over-dependence on the former Soviet Union for technology introduction and implementation ultimately proved to have a negative impact when Moscow suddenly withdrew its experts and terminated all assistance in 1960 due to rising political tensions between the Communist Party organizations in the two countries.

Several agreements were critical in terms of laying the initial overall framework for S&T cooperation between Moscow and Beijing. The 1954 Sino–Soviet Agreement on Science and Technology Cooperation facilitated Moscow's 156 technical aid projects, mostly in industrial production and equipment, as well as the establishment of a special joint committee that administered and managed S&T cooperation between the two countries (Hou 2013). Moscow provided Beijing with a significant amount of technical data and documents, such as design data for power plants, coal mines, machinery, teaching outlines, and technical standards. Most of the projects were located in China's old industrial northeast region. In 1956, Moscow sent S&T experts to Beijing to help China formulate its "12-Year Plan for Science and Technology Development," which was a milestone for setting in place China's S&T efforts under Mao Zedong. The development of the plan as well as the development of China's entire post-1949 S&T system was heavily shaped by Russian influence – and it took major reform efforts under Deng Xiaoping after 1978, lasting till the start of the twenty-first century, to come out from under the heavy weight of that Soviet influence (Lewis and Xue 1991).

Soviet assistance in S&T talent cultivation was conducted in three ways: China sent experts and outstanding S&T professionals to the former USSR, either as interns or researchers, to work and gain knowledge in areas that were seen as most urgent for economic and industrial development. These professionals would return to establish the foundation of critically needed technologies for growing the Chinese economy. In some instances, China would directly recruit Soviet experts to help set up scientific research institutes within CAS and relevant departments, and promote comprehensive cooperation with the Chinese S&T community. Large groups of Chinese professionals were organized to receive training by Soviet experts already in China to support ongoing development projects. Training in the former USSR helped to spearhead the development of China's computer industry in the 1950s; the majority of the first cadre of computer scientists in China were all trained in the former Soviet Union.

Apart from S&T support in the civilian sector, Moscow also provided technologies that were of great importance for developing China's military capability and national defense. In 1954, Khrushchev agreed to assist China in developing atomic energy for peaceful purposes, in exchange for Mao's political support; this was the first step in China's research and production effort in nuclear weapons. In 1956, the Eastern Atomic Energy Institute was established in Dubna (a designated "science town" in the Moscow *oblast*). China shouldered 20 percent of the costs for construction and operation; Moscow, 50 percent. To a certain degree, this joint endeavor helped lay the theoretical and personnel foundation for the Chinese nuclear weapons program (Lewis and Xue 1991). In 1958, a heavy water reactor, cyclotron, and a scientific nuclear research facility were completed in Tuoli, a suburb southwest of Beijing, which enormously improved research conditions for China's nuclear physics program. Sophisticated technology and equipment were provided to support the research, design, and production of China's first atomic bomb and missile delivery systems.

Needless to say, this brief period of close China–Soviet technology cooperation reflected Moscow's *realpolitik* considerations, even if much of it was couched in terms of a Communist brotherhood. After Stalin's death, Khrushchev's decision to assist China develop nuclear energy for peaceful purposes occurred within the context of an intense power struggle in which Mao's support was critical for the eventual victor. Moscow's support for China's nuclear program subsequently expanded to include weapons-related technologies in 1957, after Mao expressed his support for Khrushchev, who was under threat of being overthrown in Moscow by a conservative group who objected to his program of "de-Stalinization" (Shen and Xia 2012, 2015). Mao, however, soon became disenchanted with Khrushchev's de-Stalinization campaign and let his dissatisfaction be known. Predictably, the flow of Soviet nuclear aid to China became increasingly limited in pace, scope, and depth when Khrushchev's position was firmly secure (Shen and Xia 2012). In addition, as China fell into the turbulence and radicalism of the Cultural Revolution (1966–76), the Russians had become increasingly concerned about what was happening in China, about Mao's leadership, and about security issues along the Sino–Soviet border.

Phase Two saw cooperation between the two countries come to a grinding halt. Military tensions about border issues along the Amur (Ussuri) River on the Chinese northeastern border as well as the revolutionary posture of the Chinese Communist Party in its relations abroad made for difficult times. It was not until Gorbachev came to power and offered an olive branch to China that S&T cooperation could be restored. Gorbachev offered to work with China to build a railroad line linking Urumqi and Kazakhstan, to engage China in Russia's space program, and to resolve the navigation channel issue on the Amur River (Wilson 2004).

In Phase Three, the so-called "post-Cold War era," China–Russia S&T cooperation has become increasingly strengthened and institutionalized, the result of both traditional political ties and the practical need to maintain strategic coordination to balance the power and influence of the United States. Shortly after the collapse of the USSR, Beijing sent a vice minister level S&T delegation to Moscow to establish inter-governmental S&T relations. In 1992, the two sides concluded the Agreement on China–Russia Science and Technology Cooperation, setting up the Standing Committee for S&T cooperation at the vice premier level under the Sino–Russia Committee of Economic, Trade, and S&T Cooperation. More than 200 inter-government programs were formulated during 1993–6, covering almost all aspects of S&T development. The mechanism of regular meetings between Chinese and Russian premiers was established in 1997, which was a historic milestone in the process of bilateral S&T relations institutionalization.

The 1998–2012 period can be categorized as a time of exploration for high-tech industry transformation and innovation cooperation. China and Russia signed a Memorandum of Understanding (MOU) for Innovation Cooperation, creating a working group to guide, super-vise, and facilitate joint R&D in such diverse areas as nuclear energy, telecommunications, shipbuilding, environmental protection, biotech, aeronautics, and astronautics. The Sino–Russian Science and Technology Park in Changchun began operation in 2006 as a demonstration project for cooperation in wider areas.

From 2012 onward, Sino–Russia S&T cooperation gradually has shifted from short-term, small-scale to mid- to long-term cooperation on big projects. An MOU was initiated to direct joint efforts in priority areas including nanotechnology, material science, life science, energy, and information and communication technology. The most recent important development is the first Sino–Russia Innovation Dialogue convened by MOST and Russia's Department of Economic Development in June 2017. The dialogue engages some 200 representatives from government, universities, research institutes, industry, investment institutions, technology transfer institutes, and high-tech innovation enterprises. The two parties issued a joint statement that commits bilateral concerted efforts to coordinate national innovation policies and strengthen communications over issues such as innovation strategy, trends, construction of national innovation systems, technology transfer, mass entrepreneurship, S&T finance, and industry conglomeration. In addition, China and Russia will support cooperation between business incubators from both countries, encourage young people to start their own businesses, strengthen cooperation between Chinese and Russian science parks, and push for the establishment of a China–Russia technology industry cooperation platform.

China–Japan S&T relations

S&T exchanges between China and Japan began in the 1960s, initially conducted largely by civil society organizations, with limited government participation.[6] In 1978, following the normalization of relations six years earlier, the Japanese government established official cooperative S&T links with China; the principal participants were Japan's Ministry of Foreign Affairs (MOFA) and China's State S&T Commission (later Ministry of Science and Technology). Cooperation during this period was characterized largely by one-way technology transfer to China, which was then eager for the scientific knowledge and industrial technology it regarded as indispensable for building its basic science and research system and industrial base. The 1980 Agreement on China–Japan Science and Technology Cooperation marked the inception of the so-called "horizontal" cooperative mechanism that expedited cooperation and, more importantly, significantly expanded the forms, channels, and participants involved in the cooperative S&T relationship.

Despite the often-strained state of the bilateral relationship stemming from the unresolved issues associated with World War II, Sino–Japan S&T cooperation is increasing (Yahuda 2013). Expanded cooperation is being conducted under the overall framework of several important agreements, including Agreements on China–Japan Science and Technology Cooperation, China–Japan Cooperation in Environmental Science, and China–Japan Nuclear Energy Cooperation; exchanges and cooperation through the Japan International Cooperation Agency (JICA); and direct cooperation between the S&T ministries and departments of each country. Personnel exchanges also are witnessing rapid increase, in that major Chinese government departments and research institutes have established regular and stable cooperative partnerships with Japanese counterparts.

The Joint Committee on Sino–Japan S&T Cooperation serves as an important organization that oversees, administers, and promotes exchanges and joint R&D programs. The 10th annual meeting held in Tokyo in 2003 was of particular significance in that both parties pledged increased collaboration based on the principle of "equal status and mutual benefits." There was an emphasis on high-level exchanges, encouraging the active participation of universities, research institutes, and industries. China and Japan agreed that the focus of cooperation in the future should be on biotech, life sciences (including agricultural and food technologies), IT, nanotechnology, energy, and the environment. The latter has arguably proven to be the most effective, given its large scale and high level of personnel exchange, covering wide areas of cooperation. Beijing and Tokyo signed the first agreement on environmental protection in 1994, and the inter-governmental joint committee organized the first conference to designate a series of environmental protection programs. In 1996 Japanese Prime Minister Takeshita Noboru initiated the China–Japan Friendship Environmental Protection Center through Japan's Office of Development Assistance (ODA). Currently, the center plays an important role in pollution prevention technology, environmental monitoring, environmental information, environmental strategy and policy studies, personnel training, and environmental technology exchanges. Japan has been particularly concerned about the level of acid rain flowing across Northern Japan from the industrial pollution in China's northeast – where many older factories still continue to produce goods using dated technology and energy sources, including burning China's notoriously dirty coal. Japan also is concerned about the waters off its coastal areas, given China's offshore drilling activity and the extensive Chinese fishing fleet in the area.

Apart from inter-governmental cooperation, non-governmental S&T exchange and cooperation also are playing an increasingly important role, as investments and R&D centers established by Japanese

---

Box 4. Internet of Things

The "Internet of Things" (IoT), under which everyday objects (from kitchen appliances to cars to wearables) are connected to the Internet, enables embedded sensors to collect and exchange data. This, in turn, creates opportunities for consumers, businesses, and governments, as more and more devices are connected to the Internet and thereby each other. IoT gives homeowners the ability to control the temperature within their houses, water their plants based on growing needs and environmental conditions, and turn on or off any home appliance, all while away on vacation. Businesses can use IoT to optimize operations, boost productivity, and reduce resource use by keeping better track of assets, equipment, and consumer preferences. Cities can benefit from IoT to optimize traffic flow, schedule trash pickup based on need, or minimize streetlight use based on time of day, season, and weather conditions. The number of such connected devices worldwide is growing annually, with 8.4 billion estimated devices in 2017, an approximately 30 percent increase from 2016, and a projected 20.4 billion devices by 2020 (Gartner 2017).

China is keen to become a global leader in IoT, and initially identified it as a key field for R&D in its 12th Five-Year Plan. In its 13th Five-Year Plan, China emphasized the importance of integrating Internet-based innovations with traditional sectors, such as health, education, and transportation, as an important source of economic growth in an initiative called Internet Plus. IoT – along with cloud storage, big data, and cloud computing – is a key component in the Internet Plus initiative (Koleski 2017). With strong governmental backing, and less public concern and say-so regarding privacy protection, China has the ability to move faster than its Western counterparts in the implementation and adoption of IoT. In 2016, China's IoT market was estimated at approximately 900 billion yuan (US$135.5 billion) and is projected to reach 1.5 trillion yuan (US$226 billion) in 2020 (*China Daily* 2017b). A variety of IoT applications already have been implemented throughout China, including smart parking, smart water meters, and monitoring of environmental pollutants (Guerrini 2016). Chinese IoT companies also are working on IoT applications to secure communications, improve medical monitoring of patients, track assets, manage fleets, and more (Chen et al. 2014).

In Shanghai in 2016, Huawei and China Unicom jointly released a smart parking solution based on NarrowBand-Internet of Things (NB-IoT), allowing drivers to find available parking spaces and relieve traffic congestion (Huawei 2016). NB-IoT is considered a promising IoT Low-Power Wide-Area Network (LPWAN) option for "things" such as sensors and connected devices to communicate with one another, as it uses lower power and bit rates compared to traditional wireless networks. Huawei is working to expand the use of NB-IoT to other applications such as smart metering and logistics asset tracking (Huawei 2016). Meanwhile, US and European telecommunication giants such as Verizon Communications, T-Mobile, Sprint, Qualcomm, and Vodafone also are investing in NB-IoT with services predicted to roll out in the US and certain European countries in 2018 (Qualcomm 2017; Vodafone 2017; Zacks Equity Research 2017).

high-tech enterprises are rapidly growing. Demonstration projects have included Sharp Wuxi (LCD), SGNEC (chips), Shanghai Huahong NEC (semi-conductors), Shanghai Fanuc (robots), world telecommunications tycoon NTT Docomo (Internet), and Huawei and China Unicom (Internet of Things).

Japanese enterprises are seeking greater cooperation with Chinese universities to expand their business channels in China. For instance, Sumitomo and Shanghai Jiaotong University signed an agreement to foster joint R&D, personnel training, and co-funded technology development programs with high potential. The establishment of the Daikin–Tsinghua R&D Center marked the first S&T initiative in China by Japanese air-conditioner makers, intended to develop energy-saving technologies. Other Japanese industrial leaders also have established overseas R&D centers in China, including Toshiba, Ricoh, Fujitsu (Embassy of PRC in Japan 2007). In addition, with the platform provided by MOST and the Japan Science and Technology Agency (JST), universities in both countries are able to cooperate on S&T innovation and other urgent S&T-related issues. In 2016, MOST and JST initiated a joint program on urban environment and energy with participation from Chinese universities (Tsinghua, Beijing, and Zhejiang) and Japanese universities including the University of Tokyo, Tohoku University, and Nagoya University (Embassy of PRC in Japan 2016).

Notwithstanding these increases in Sino–Japan S&T relations, they occur within an overall framework of the political strains mentioned earlier stemming from historical conflicts, current territorial disputes, and worsening security competition in East Asia (Newby 2018). In response to a unilateral move by Tokyo to nationalize the Senkaku/Diaoyu Islands in 2012, Beijing canceled the annual meeting organized by the China–Japan S&T Cooperation Committee, which wasn't resumed until 2015. A recent highlight of the new engagement between the two countries involves energy conservation and environmental protection. The 2nd Sino–Japan Energy Saving and Environmental

Protection Science and Technology Summit was held in Dongguan in December 2017, which facilitated the confirmation of multiple projects between Chinese and Japanese enterprises in energy saving and air pollution control.

The first Sino–Japanese Energy Conservation and Environmental Protection Technology Summit forum was held in Dongguan in 2016. This forum, along with the China–Japan Energy Conservation and Environmental Protection Cooperation Pavilion, have been established as the permanent activities of China (Dongguan) International Scientific and Technological Cooperation Week (Wen 2017). China's energy-saving and environmental protection industry is developing rapidly with huge investment and a vast potential market; Japan has advanced technology and management experience that China requires (Swanstrom and Kokubun 2012). In 2017, the Guangdong Provincial Department of Science and Technology also released the Guidelines for Joint Innovation International Cooperation Projects, focusing on encouraging projects jointly supported by China and Japan in various fields for the purpose of moving new ideas into commercial production.

China's S&T relations with the EU

S&T relations between China and the EU have undergone fast development since the normalization of diplomatic ties in 1975.[7] The agreements between the EU and China exist in parallel with a host of bilateral S&T agreements that China now has in place with various EU members. With Brexit and the departure of the UK from the European Union, the Sino–UK S&T relationship will take on added importance for the two countries. The EU and China signed a formal Science & Technology Cooperation Agreement in 1999, implemented through a joint steering committee, which has since provided guidelines and an overall framework for cooperation. In 2008, the European Atomic Energy Community and the Chinese government signed an agreement that put in place R&D cooperation for peaceful uses of nuclear energy (Euratom 2008). In 2003, the EU–China Comprehensive Strategic Partnership was created and cooperation in a wide range of areas has been deepened and broadened, resulting in high interdependence today. Both parties adopted the EU–China 2020 Strategic Agenda for Cooperation and had the first High Level Innovation Cooperation Dialogue during the 16th EU–China Summit held in November 2013. Through regular meetings and a broad range of sectoral dialogues, the Strategic Agenda has been implemented under the cooperative umbrella set by the annual High Level Strategic Dialogue, the annual High Level Economic and Trade Dialogue, as well as the bi-annual People-to-People Dialogue.

China has been recognized by the EU as a key partner in science, technology, and innovation, with EU–China cooperation intensifying in recent years (EU 2015; Le Corre and Sepulchre 2016). China has been the third most important international partner country under the Framework Programme 7 (FP7) that ran from 2007 to 2013. According to a report on China–EU Research and Innovation Relations, with 383 participations of Chinese organizations in 274 collaborative research projects and a total EU contribution of €35.24 million, China remains a significant player in Horizon 2020 (H2020), the EU's special Framework Programme for Research and Innovation, running from 2014 to 2020. So far, 227 applications from China resulted in 187 eligible proposals, with 60 participations from Chinese organizations in 33 main listed projects. Moreover, the well-recognized Marie Skłodowska-Curie program included around 959 Chinese participations (EU 2015).[8]

Among all EU Member States, China's S&T relations with Germany are perhaps the most stable and productive (Shambaugh and Sandschneider 2007). Germany has traditionally loomed large in the Chinese perception of the world S&T landscape due to Germany's strong industrial competitiveness and R&D capabilities. During Premier Li Keqiang's 2017 visit to Germany, the two parties announced a "Plan of Action for Sino–German Cooperation: Shaping Innovation," which provides for a strategic high-tech project "Industry 4.0," a German initiative, and programs on urbanization, industrialization, informatization and agricultural modernization, which are China's policy priorities. This is likely to result in increasing complementarity and coordination between "Made in China 2025" and "Industry 4.0," facilitating innovation and global standard setting in the field of smart manufacturing. Germany's Federal Ministry of Education and Research (BMBF) issued its China Strategy in 2015, and China's MOST issued "Jointly Shaping the Future through Technology Innovation: Germany Strategy" in 2016, reflecting consensus on a shared responsibility to lead a new round of innovative industrial and economic change – one based on increased policy dialogue and enhanced S&T cooperation between the two countries.

There remain, however, some areas in which Germany shows little enthusiasm to cooperate, out of deep-seated concern that cooperation in some high-tech fields (e.g., development of new automobile engines and solar panels) will erode its technologically competitive edge. The German government remains cautious in its approach to cooperation with China, given China's poor record regarding IPR protection (European Commission 2018).

FOREIGN R&D IN CHINA

In the late 1970s, as part of the so-called "new international economic order" (Bhagwati 1977), many multinational firms set up R&D centers in developing countries as a way to exhibit their commitment to technology transfer and to Third World economic development. In the majority of cases, however, these R&D centers were largely hollow operations, with little of substance – research or training – taking place inside, except for some local product adaptation. Today, we see an opposite picture emerging in the case of China and other BRICs countries. While clearly not all of the more than 1,800 foreign R&D

centers now located in China are engaged in state-of-the-art research – basic or applied – and most have eschewed a focus on basic research, there are a growing percentage of foreign companies who are filling out their complete value chain in China by deepening their R&D activities as part of a strategic global repositioning of their business (Sun et al. 2013). An important example involves Apple Corporation, which in late 2017 announced its decision to launch two new R&D centers in China, one in Shanghai and one in Suzhou. Another involves Google, which in early 2018 announced its intention to build a new China-based research center focused on artificial intelligence. These decisions reflect a strategic decision to be in closer alignment with PRC S&T priorities and technology targets.

Like many other critical transitions in China's economic moderniza-tion drive and its relationship with the outside world, the growing role of foreign R&D in the PRC is being driven by a confluence of govern-ment and market forces (Sun et al. 2008). First, as indicated earlier in this book, the Chinese government has emphasized the importance of strengthening the country's technology base and upgrading the innova-tive potential of PRC enterprises (Zhou et al. 2016). Accordingly, in April 2000, the Ministry of Foreign Trade and Economic Coopera-tion (MOFTEC, now the Ministry of Commerce, MOFCOM), issued Circular #218, which basically formalized the status of foreign R&D centers in China by providing guidance and details on the rules for their establishment. In April 2002, MOFTEC's foreign investment legislation was modified to change R&D activity from a "permitted" to an "encouraged" form of foreign investment. At the time, these policies complemented a series of related changes that have taken place with regard to the importation of foreign technology. Moving away from the restrictive regulatory regime of the 1980s and 1990s, in 2002, Beijing radically revised the existing legislation regarding foreign technology imports. In essence, the spirit and intent of these revisions has been to promote smoother and faster movement of tech-nology and know-how into China by shifting the PRC government emphasis toward approval rather than tight control. This may seem ironic given foreign corporate concerns about excessive pressures on them to engage in technology transfer activities in a more concerted manner, but it is consistent with Chinese thinking that bureaucratic obstacles should not constrain the flow of technology from abroad (Walsh 2003).

Under the rules for foreign R&D centers, ownership structures can vary from equity joint ventures to wholly owned enterprises. To qualify for formal R&D status, however, 80 percent of the staff must hold at least a college degree and be involved in actual R&D activities. Two types of R&D activities are permitted under the legislation: An R&D center whose main purpose is to engage in the general transfer of know-how to any entity; and an R&D center that is controlled by a parent firm and is involved in research for which it will be reimbursed expenses plus a reasonable profit. In the latter case, the expectation is that the IPR belongs to the parent sponsor. R&D centers, however, cannot engage in so-called "technology trade" that is not the product of their own R&D efforts. These foreign R&D centers remain eligible for a range of tax incentives as well as tax relief for equipment imported to support the R&D activities. In addition, the Chinese government has committed itself to easing visa requirements to enable entry and exit to/from China for both locals and foreign nationals called "high end talent" employed at the center.

Moving beyond the preferences offered by the central government, both Beijing and Shanghai have issued their own regulations to further encourage foreign companies to set up R&D operations in their respective cities. Some of these regulations are aimed at attracting expertise from outside China's coastal areas by, for example, awarding residency permits. In October 2017, Shanghai authorities issued a series of new measures to attract more foreign-funded R&D centers to the city as part of the efforts to transform it into a global technological innovation hub by 2030 (Zhuo 2017). According to the 15-article policy document approved by the Shanghai municipal government, high-level foreign R&D centers in Shanghai are now granted the same privileges as regional headquarters of multinational companies, including faster and smoother exit and entry processing and more simplified customs reviews. As of the end of 2017, there were approximately 425 MNC R&D centers in Shanghai, involving well over 40,000 employees.

A second driver behind the growth of foreign R&D centers in China revolves around the issue of technical standards. Since the mid-1990s, Chinese government policy has placed a greater emphasis on acquiring technical know-how to enable local industry to gain a greater percentage of the revenues associated with licensing and technical standards. In January 2005, MOST's Development and Planning Department decided to provide a new injection of funds into R&D for the purpose of establishing 29 international technical standards. The original program, which began in 2002, now involves more than 2,500 scientists and experts working in such fields as environmental protection indicators, trace element examination, textile safety, broadband local area networks, and radio-frequency identification (RFID). Given the steadily expanding size of the Chinese domestic market and the potential weight of Chinese market power on an international level, foreign firms have been anxious to shape or influence China's decisions regarding which standards are being adopted in such areas as telecommunications, software, computers, and pharmaceuticals. Perhaps no area better illustrates how the competition for standards setting has drawn in foreign R&D investment than mobile telephony. Siemens, Nokia, Ericsson, and Motorola all made substantial investments in building out R&D operations in China in this highly competitive sector. As the requirements and sophistication of Chinese consumers continue to rise in the highly dynamic mobile phone market, especially as the so-called "shared economy" and "cashless economy" both grow, more local R&D will be needed to get new products into the market quickly and reliably, thus helping to set trends and win market share.

Cost-cutting considerations are clearly a third driver for attracting foreign R&D to China. The data seem to vary from city to city and from province to province, but the fact remains that the loaded costs of employing and supporting an engineer in China are nearly the same as hiring a US counterpart. The once attractive pathway to cost-based substitution in high-end engineering is steadily disappearing; the new critical issue is talent availability. In many instances, the movement of R&D to China by foreign firms reflects a desire to create a critical mass of talent, at affordable rates, that can be utilized to focus on a core problem or alternative technical solution that otherwise might be ignored and bypassed due to lack of available staff and funds in the US. Moreover, the presence of an advanced technical team in China, especially with local language skills and cultural familiarity, gives the foreign firm a better chance to work with local suppliers and vendors to ensure that domestically manufactured parts and components meet required levels of quality and performance. Once local Chinese R&D teams can be integrated culturally and operationally within the global R&D infrastructure of a large multinational firm, they are ready to service global markets as well as the local Chinese marketplace. This clearly has been the intention of firms such as Microsoft and IBM – both of whom have steadily grown their research presence in China (Grimes and Miozzo 2015).

Professional services companies in the human resources field, more commonly known as "headhunters," have found that the demand for their services has substantially increased over the last several years. In the 1980s and 1990s, the major headhunters, mostly based out of Hong Kong and Singapore, spent the bulk of their time finding appropriate expatriates to take top managerial assignments in China. Today, they have expanded their operations to Beijing and Shanghai, and their principal focus is largely on identifying experienced PRC nationals in China and abroad who wish to return home to assume a leadership role in these types of foreign-invested R&D centers and technical organizations. Chinese scientists and engineers, at home and abroad, are drawn to working in foreign R&D organizations because of the nature of the projects, the opportunities for training and travel overseas, better salaries (though not always), and more varied career opportunities. As the staffing needs for these foreign R&D operations have grown, the result has been the creation of an emerging "internal brain drain" problem, with some of the best and brightest Chinese talent forsaking opportunities with domestic companies and government labs for the seemingly more exciting career path in foreign-invested organizations.

Fortunately – or unfortunately, depending on one's perspective – this problem may be short-circuited by the further growth of technological entrepreneurship in China. There is a saying in Chinese, "it is better to be the head of a chicken than the tail of an ox" (*ning wei ji tou, bu wei niu wei*). The high turnover rate for junior- and mid-level talent in both foreign-invested and domestic R&D operations reflects their apparent willingness to further "jump into the sea" and embark down an entrepreneurial path that increasingly involves starting their own firms. This is not much different than what happened in Taiwan in the late 1970s and 1980s in Hsinchu Science Park, when many local engineers left employment with foreign companies such as Motorola and General Instruments to open their own firms – sometimes with indirect government support and even encouragement. One particular difficulty that already has arisen in China from the rapid circulation of such technical talent, however, concerns the security of IPR and adherence to confidentiality agreements contained in employment contracts with their foreign employers. With many foreign firms utilizing trade secrets and not always patents to protect their IPR, it is sometimes hard to prevent critical know-how from being used inappropriately in some of these startup firms. This also is the case with some returnees from abroad, who have left positions with US-based firms to begin an entrepreneurial journey in China.

There are a range of other drivers that account for the step up in the number of foreign R&D centers being established in China. Some of these factors exist on the "push" side rather than the pull dimension. They include tax and visa policies at home in the US and Europe, the growing pressures on compensation and benefits packages, and overall problems regarding the availability of well-trained technically oriented individuals. Most critical, however, remains the imperative of global competition, which continues to be creating more pressures for more sustained innovation, greater customer responsiveness, and more rapid commercialization of new products and services. China's role in this regard promises to be anything but passive, especially as it seeks to secure for itself a more prominent place in the global value chain. PRC government policies are distinctly based on the notion that the expanding number of foreign R&D centers will serve as a catalyst for sparking new innovative behavior throughout the economy and attracting more global talent.

It is safe to say that the contributions from foreign R&D activities in China still remain limited, though this has much to do with the fact that the phenomenon is still in its initial stage of development. A number of important questions remain, nonetheless. For example, will foreign R&D in China become an even more integral part of the PRC's national innovation system? Is there a formal capture strategy in place or being conceived by the Chinese government to ensure that the contributions from R&D can be absorbed? And, is China's national innovation system structured and developed to the point that it can maximize the benefits from being steadily embedded in a comprehensive web of global knowledge networks in world science and international engineering? At the present time, the response to these three key questions would seem to be, "stand by, the answer is yet to be fully determined." That said, from both policy and organizational perspectives, there has been growing evidence that the Chinese S&T system is indeed pointed in the right direction as it seeks to optimize the growing presence of foreign R&D activity (Zhou et al. 2016). While the direction of China's technological progress may not always be linear, aided and abetted by the development of continuously more cohesive relationships with the world's leading technology-based corporations, the pace of progress will likely be more rapid than we might anticipate.

To be more specific, so far, the identifiable contributions from foreign R&D in China seem to lie more in the world of intangible benefits rather than concrete ones. Nonetheless, they still are critically important as a precursor to more rapid Chinese technological advance. They fall into the following areas, many of which in the past have been areas of major weakness for the PRC:

+ Training: technical training, cross-functional/cross-cultural teaming, and product and process design methodology, especially electronic design automation for accelerated design cycles;
+ Technology transfer: the diffusion of "uncodified" trade secrets rather than specific patented information;
+ Standards: best practices, industry standards, performance metrics, and quality requirements;
+ Software: programming methodologies, software design architectures, systems, integration techniques, and overall testing procedures and quality assurance;
+ Management: project management, business management, and management of knowledge workers;
+ Networks and information resources: participation in global knowledge networks;
+ Spin-offs: new business ventures and entrepreneurial activity; and
+ Spillovers: technical assistance to vendors and suppliers.

In the final analysis, however, Chinese policymakers fully recognize that the R&D activities of foreign firms in China are driven largely by the strategic agendas of these companies. To gain a deeper, longer-term commitment from foreign firms in the R&D area, China will have to improve its overall enforcement of IPR protection as well as its business environment. This also is true with respect to China's efforts to develop its advanced software industry, especially if the country hopes to move beyond basic outsourcing activities. The need for better IPR enforcement is often affirmed by many academic, business, and legal observers of the Chinese scene, though with little expectation that much will be done in the short term. Strong IPR enforcement also is necessary for MNCs to be willing to engage in more extensive basic research in China. Securing this type of scientific-oriented research is very much coveted by China's S&T leadership. Venture capital will be hesitant about supporting technological entrepreneurship if there continues to be pervasive apprehension that IP rights cannot be made secure.

Based on the experience of other Pacific Rim economies, the key to solving the IPR problem in China actually lies in the degree to which the roots of local technological entrepreneurship take hold. With locally created IP at risk, the appropriate conditions now clearly exist for local government and enterprise stakeholders to make progress in cracking down on those who violate foreign and domestic IP rights. Recent progress in IPR protection plus more internal security controls by these foreign firms may produce a more ideal setting for further expansion of serious foreign R&D initiatives in China during the next 5–10 years.

MAIN OUTSTANDING ISSUES AND CHALLENGES

In spite of the overall progress China has made in institutionalizing its international S&T cooperation structure and expanding its cross-border S&T relationships, numerous challenges remain. IPR protection has been, and will continue to be, a serious concern for foreign S&T partners – in both public and private sectors. The rise of China as a more active player in global S&T affairs has reflected its strengthened S&T capabilities, thus reducing the S&T gap with developed countries and shifting its relative position from a poor underdeveloped country to an emerging technological superpower. This transition has significant implications for its S&T cooperation efforts. Technology imports shaped much of China's cooperative relations during the time when China was playing catch-up; many foreign firms were willing to indulge China even with its lax IP protections as the price of gaining entry to the world's largest and fast-growing market. Now that Xi's Chinese dream is being realized, and China increasingly is viewed as a serious competitor, relations have become more difficult across a broad spectrum of issue areas. For example, given China's plans for massive investments in development of artificial intelligence, will Western countries be willing to collaborate with China and perhaps put their technology at risk? Along with the rise of China's position across the global innovation landscape, it has become increasingly difficult for China to play the role of learner in its cooperation with developed countries. Clearly, China is in the process of redefining its role – one that necessarily requires a more co-equal partnership in terms of cooperation and contribution. This will require China to afford far greater IP protection for foreign partnerships; at present, Patent Cooperation Treaty (PCT) applications by China are roughly one-third that of the US.[9]

Despite these challenges, some appreciable progress is being made. The US–China Clean Energy Research Center (CERC) provides one illustrative example. CERC is characterized by public–private consortia underpinned by a strong IPR protection agreement. A special IPR Annex is part of the founding protocol. According to CERC's 2012–13 annual report, projects under the Advanced Coal Technology Consortium yielded 17 patents, and projects under the Clean Vehicles Consortium resulted in 20 patents and invention disclosures in the US and 12 patents in China (US–China Clean Energy Research Center 2013). China's diminishing asymmetry also opens up broad new avenues for substantive bilateral and multilateral cooperation, as China becomes a more important contributor to the world's S&T literature, producing a growing number of top-tier cited refereed articles (Suttmeier and Cao 2006). The European Union and the Chinese government, for example, have recently agreed to set up a joint project funding mechanism involving annual investments of roughly €100 million and 200 million yuan (US$200 billion) in support of joint projects between EU and Chinese agencies.

Over the past four decades, China has achieved significant gains in international S&T cooperation, spurred on by rapid economic development and its opening-up policy. China now sees international S&T cooperation as part of a new stage in its S&T development, in which there will be greater demand for international S&T cooperation at all levels and among public and private stakeholders. Along with China's improvement in its S&T capacity and core competencies, China's role in international S&T cooperation is changing gradually from learner to partner and rule maker. We expect to see increasing proactive participation by China in global S&T governance, as Chinese scientists hold a growing number of positions at major international S&T organizations, and more Chinese-initiated "big science" projects and advanced research facilities that attract scientists from all over the world.[10]

Under the new round of reforms launched by the 13th Five-Year STI Plan and Strategy of Innovation-Driven Development, China has put forth a strategic vision for future international S&T cooperation that includes very ambitious goals and innovative mechanisms. If reforms are successfully implemented, they should increase the openness of China's S&T programs, resulting in growing demand for international cooperation. Through comprehensive reforms, some of the issues that have thus far hindered S&T cooperation, such as restrictions on travel abroad and the use of funds, might be resolved. We expect that through new reforms, some of China's challenges in large part can be overcome.

Nonetheless, the Chinese government needs a clearer definition of its key role – one that improves the quality of its services to China's major innovation actors. It already is reinforcing its international S&T cooperation strategy through such efforts as promoting innovation dialogues, expanding cultural and educational exchanges, upgrading the scale of communications, and involving an expanded number of stakeholders such as universities, research institutes, and private enterprises. The government also is setting up special funds and programs, with different purposes and characteristics, to promote international S&T cooperation. More resources are being channeled and leveraged from central and local government, as well as the growing private sector. In the long run, China needs to develop a strategic plan and policy umbrella that will better guide its international cooperation activities and design more innovative mechanisms to better meet the country's changing needs. It clearly is an appropriate time to introduce additional reforms that will foster mutually beneficial international S&T cooperation; these reforms will have to provide more incentives to potential and existing foreign partners that will overcome the anxieties and uncertainties that up to now too often have constrained the growth of new activities.

The bottom line looking ahead is a simple one – there is no international S&T-related issue whose solution will not require close cooperation and collaboration with China. Climate change, clean energy, global pandemics, water, and other such issues are central to China's future and mission critical for the world if it is to avoid major disasters in the coming years. China's decision in 2017 to step up on global climate change despite the US decision to withdraw from the Paris Accord signed during the Obama Administration marks an important turning point in China's role in the international S&T system. Simply stated, China's willingness to take on a leadership role in this issue portends an expanded Chinese presence across multiple similar issue areas. Chinese behavior is starting to re-shape the global S&T and innovation landscape. How countries such as the US and Japan as well as the EU will deal with this new Chinese posture remains one of the key challenges facing the international S&T system.

During the 19th CCP National Congress held in October 2017, despite the sense among many foreign observers, the party's General Secretary

Xi Jinping indicated that China will continue to attach great importance to openness; Xi asserted that openness is critical for turning China into a true innovative country with global competitiveness. Under Xi's leadership, in spite of an obvious increase in nationalist spirit, China promises to become more and more open, will combine "bring in" and "go global," give priority to promoting the Belt and Road Initiative, and strengthen international cooperation to enhance its innovation capacity.

To achieve its goals, Beijing intends to make use of S&T comprehensively to advance major power diplomacy through strengthening and refining top-down designs for international S&T cooperation, deepening and expanding innovation dialogue mechanisms with major countries and S&T partnerships with developing countries, proactively initiating and coordinating international big science projects and programs, and attracting high-end overseas S&T talent. There is a growing realization among PRC leaders that China is rapidly moving toward the center of the global innovation stage, becoming a leader in a number of important fields, and shifting from being a passive follower to *"san pao bing cun"* (coexistence of catching up, running neck and neck, becoming top runner). Domestically, the major challenge facing Chinese society is people's increasing demand for high-quality life versus unbalanced, insufficient development. By pressing harder to enhance the performance of the research sector, the leadership hopes that advances in its S&T innovation capabilities can offset current shortcomings facing the Chinese economy.

At the so-called "Liang Hui" or two congresses held in March 2018, Chinese policy regarding S&T and innovation appears to have undergone even further changes. China's strategic high technologies are increasingly approaching the global frontier; the PRC has entered an historic stage of "running neck and neck, becoming top runner" while being in less of a "catching up" mode. Before his retirement as MOST minister, Wan Gang urged the Chinese S&T community to strengthen openness and cooperation so as to proactively take part in international innovation and entrepreneurship as well as more efficiently leverage innovation resources both at home and abroad. According to Wan (2018), China's S&T diplomacy will be further enhanced by creating new dialogue mechanisms within existing multilateral organizations such as the BRI summit, G20, and BRICS, expanding the country's S&T partnership network and diversifying prevailing methods of cooperation. The world's most advanced major S&T powers still will loom large in China's overall international S&T networks, especially when it comes to strategic emerging technologies like artificial intelligence and clean energy automobiles. Along with to upgrading the level of "mass entrepreneurship and mass innovation," the PRC government intends to enhance existing talent poticies to grow green channels for foreign experts to secure jobs in China, and attract Chinese international students to engage in innovation-related entrepreneurial activities (K. Li 2018).

According to the 13th Five-Year Plan on STI International Cooperation (China STI 2017), over the next decade China's international S&T cooperation efforts will be pushed along the following areas:

+ Create a diverse, highly differentiated S&T international cooperation framework based on Beijing's categorization of developed countries, emerging economies, and developing countries;
+ Achieve full mobilization and deployment of resources covering basic and frontier research, key generic technologies and application by strengthening inter-governmental S&T cooperation featuring bilateral and multilateral agreements, consensus, and commitments;
+ Giving greater focus to connecting the S&T innovation needs of China and its neighboring countries and the developing world; and
+ Ensuring that the BRI initiative will be characterized by more S&T-related personnel exchanges, communication and coordination of S&T policies, thus creating a China-centered hub-and-spoke system.

Beijing has announced plans to co-build cooperation platforms with BRI countries such as national laboratories and research centers, technology transfer centers, and technology demonstration and promotion bases. China has committed itself to building up the S&T capacity of developing countries both in hardware (research facilities) and software (knowledge and talent pool). This includes encouraging and supporting foreign scientists to initiate and participate in strategic research, and formulation, implementation, and evaluation of guidelines as well as strengthening the local talent pool to meet the demands of the new economic situation. Based on comments in the Chinese media as well as in the speeches of PRC officials, the government is determined to expand the channels for talent introduction, attract more high-end overseas Chinese and foreign experts, and promote Chinese scientists to high positions in international S&T organizations. This may help to explain why the former State Administration for Foreign Experts is being incorporated into the MOST organization.

Equally important, Beijing has suggested enterprises also will play a more active role in promoting the country's international S&T innovation cooperation. They will be absorbed into inter-governmental S&T cooperation mechanisms, and those in good financial condition will be supported to establish overseas R&D centers to carry out international industry–university–research institute cooperation. Also, foreign companies will continue to be encouraged by the Chinese government to set up R&D centers and labs in China. Looking ahead, given that the country aims to deepen engagement in global S&T innovation governance, we likely will see more Chinese efforts at agenda setting for the global innovation system and more emphasis on rule setting for key international S&T projects to address key global challenges including food security, energy security, environmental protection, climate change, and public health. It remains unclear, however, whether the international S&T community will welcome a stepped-up Chinese presence without a series of concomitant gestures from Beijing with respect to prevailing norms and values in areas such as internet freedom, cyber security, IPR protection, and research ethics.

How Effective Is China's State-Led Approach to High-Tech Development?

As we have seen, the Chinese state plays a central role in fostering high-tech development. In their extensive review of efforts to explain China's approach to economic development, Peck and Zhang (2013) chronicle a bewildering array of theories intended to explain its unique blend of free market capitalism and strong state control. Efforts by "Varieties of Capitalism" (VoC) scholars to locate China somewhere on a continuum ranging from liberal market economies (e.g. the US) to coordinated market economies (e.g. Germany) ultimately fail in the case of China, since China is a hybrid with characteristics of both. On the one hand, Peck and Zhang note that China could be said to resemble a coordinated market economy, given its strong reliance on bank financing and the continuing economic dominance of SOEs.[1] State-owned banks – accounting for nearly three-quarters of all bank assets in China – rather than private lenders or venture capitalists, provide much of the financing, privileging less innovative (and CCP-connected) SOEs over private firms (Peck and Zhang 2013: 371). As a result, more innovative high-tech private firms often must rely on foreign investors for large-scale financing. Ultimately, CCP connections and highly personalized, reciprocal relationships and social networks – *guanxi* – trump the formal-legal relations said to characterize coordinated market economies. The result is an unregulated economic environment more likely to be associated with market fundamentalism. Peck and Zhang similarly argue that popular oxymoronic formulations (state capitalism, market socialism, state neoliberalism) also fail to capture the Chinese approach – as does Chalmers Johnson's notion of "plan rational," which he used to describe Japan during its developmental phase (Johnson 1982). Peck and Zhang conclude that formulations based on European and North American experiences are inappropriate when it comes to China, whatever insights they might offer. Ultimately,

> the recombinant Chinese state, which somehow holds together an unlikely marriage of entrepreneurial developmentalism with Leninist party discipline, remains the principal orchestrator of the country's development path … if this is indeed a "model," it is a complex and heterogeneous one and one that is maybe better appreciated by way of its paradoxes and contradictions than by reference to some singular logic or form of institutionalized equilibrium … the origins, the productivity and possibly even the sustainability of its "model" must be understood to be substantially anchored in its state-socialist past. Furthermore, the Chinese experience raises the question of whether it is possible for some (socialist) states to engage selectively with capitalist development, rather than buy the full "package." (2013: 385–7)

FROM INDUSTRIAL PARKS TO HIGH-TECH PARKS

If there is a Chinese approach to development, one place it is highly visible is in high-tech parks with state funding. The spatial agglomeration of economic activities has long been seen as a spur to innovation and economic development (Saxenian 1994). Geographical proximity enhances the spillover of tacit knowledge (Polanyi 1957; Howells 2002), along with more direct forms of information sharing (Audretsch and Feldman 2004; Storper and Venables 2004; Boschma 2005). Industrial parks – at least in theory – result in such agglomeration effects by actively encouraging the sharing of knowledge and innovative breakthroughs among their tenants (Bunnel 2004).

The term "industrial park" is itself misleading, since it can apply to a wide variety of spatially agglomerated economic activities. One useful typology, developed by the European Commission, distinguishes "business incubators" along two dimensions: *management support*[2] and *technology level* (each varies from low to medium to high). The resulting 3-by-3 table yields nine categories, ranging from industrial estates (low on both dimensions) to technology centers (high on both) (see figure 5.1).

In this chapter, we focus exclusively on this latter category: technology centers that are

> managed by specialised professionals, whose main aim is to increase the wealth of its community by promoting the culture of innovation and the competitiveness of its associated businesses and knowledge-based institutions. To enable these goals to be met, a Science Park stimulates and manages the flow of knowledge and technology amongst

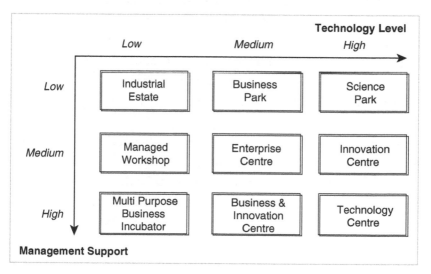

Figure 5.1 Typology of Business Incubators
Source: European Commission 2002

universities, R&D institutions, companies and markets; it facilitates the creation and growth of innovation-based companies through incubation and spin-off processes; and provides other value-added services together with high quality space and facilities. (IASP 2015)

In a review of several definitions of high-tech parks (equivalent to the European Commission's "technology centers"), Rodríguez-Pose and Hardy (2014: 17–19) identify four central characteristics that are common to these definitions. Successful high-tech parks are said to involve collaboration with major research centers and universities – the so-called "triple-helix model" involving universities, industry, and the state (Etzkowitz and Leydesdorff 2000; Leydesdorff 2013); possess a critical mass of knowledge-intensive firms to form a viable knowledge component; provide strong management support to assist with firm growth, encourage synergies, and promote technology transfer; and incubate new technology-based firms. Rodríguez-Pose and Hardy (2014) also note that high-tech parks often include some lower-tech firms, and even firms that are engaged in industrial manufacturing, since both can provide additional sources of revenue, but also can degrade the high-tech focus of a park, for example, by lowering its admission requirements.

What remains inconclusive, since existing research provides contradictory findings, are answers to such questions as: Can successful parks be created by government fiat, or are they best left to arise spontaneously, based on pre-existing local conditions? Would the high-tech firms that succeed have done so anyway even if they had not been housed within a high-tech park? Are there regional "spillover" effects, or are the benefits limited to firms within the parks? The Chinese government clearly believes the answers to these questions are affirmative, and has invested considerable public resources at the national, provincial, and local levels in various forms of industrial park development. China now has more than 1,500 national- and provincial-level industrial parks, including 168 high-tech parks (Shi et al. 2012; Moody 2015).

The Torch Program, under China's Ministry of Science and Technology (MOST), was created in 1988 to "develop high-tech industries by promoting the commercialization of S&T achievements, the industrialization of R&D results, and the internationalization of high-tech industries" (Torch Program 2011: 2). Although centrally administered under MOST, the Torch Program has largely been implemented through local experiments – a bottom-up approach aptly characterized as "experimentation under the shadow of hierarchy" (Heilmann et al. 2013). By 2016 there were 168 national high-tech parks throughout China, a three-quarters increase (from 89) over the previous five years; they accounted for approximately 12 percent of China's GDP and 18 percent of exports. According to Zhang Zhihong, director of MOST's Torch High-Tech Industrialization Development Center, "high-tech parks have become a major engine to China's economic growth. Nearly 4,300 spaces provided services for more than 120,000 enterprises in 2016, attracting investment of 5.5 billion yuan" (US$828 billion) (Xinhua 2017b; see also Torch Program 2011: 4).

CHINA'S HIGH-TECH PARKS – GATEWAYS TO INNOVATION?

Despite heavy investments into developing high-tech parks throughout China, there have only been a handful of companies that have reached the level of success that is hoped for by the Chinese government. Even among these companies, however, none have achieved the same status of international recognition enjoyed by Apple, Google, Facebook, or Amazon. The consumer base for successful Chinese companies such as Alibaba, Baidu, Huawei, Lenovo, Tencent (developer of the popular social media app WeChat), and Xiaomi is predominantly limited to the Chinese market, which, while large, does not offer the same level of prestige as the international market. China is still looking for a company that will bring the level of respect from the international community that it has been seeking since Deng Xiaoping first implemented the open-door policy. The Chinese government hopes to show the world that technology developed by China can have broader reach beyond its borders or even Asia's borders.

In this chapter, we begin with a brief history and assessment of China's first and most successful science park, Beijing's Zhongguancun, home to many of China's most successful high-tech startups. We then profile four such companies (Xiaomi, Baidu, Tencent, Didi Chuxing), highlighting the challenges they face as innovators in well-established, globally competitive technologies. We conclude by examining whether China might become a global innovator in a new, emerging technology, with a detailed case study of nanotechnology development at Suzhou Industrial Park (SIP).

ZHONGGUANCUN

In their review of dozens of high-tech parks in Latin America, South and East Asia, Africa, and the Middle East, Rodríguez-Pose and Hardy (2014) find very few countries that provide examples of parks that meet their four criteria for success. China, in their view, is one of the few exceptions, and Zhongguancun clearly meets their standards. It is located in the Haidian district of Beijing and is home to two world-class universities (Peking and Tsinghua) and more than 60 other higher education institutions, several institutes of the Chinese Academy of Sciences (CAS), 62 key national laboratories, and 251 scientific research institutions (IDG China 2015). In the early years of Zhongguancun, both before and after its 1988 designation as a science park, it was mainly known for being the best place to buy technological goods, often designed by spin-offs from nearby universities and the CAS, earning it the nickname "Electronics Street." By 2005, nearly 60 percent of all enterprises located in Zhongguancun were in information technology or related sectors (Zhou 2008). The Park is now home to more than 200 publicly listed companies, more than 200 state- and municipal-level research institutes. It has attracted dozens of the world's leading multinationals, including Microsoft, Google, and IBM (*China Daily* 2014; IDG China 2015).

A 2016 report stated that half of China's "unicorn" companies (i.e., startups that have a market value of more than US$1 billion) are located in Zhongguancun (*China Daily* 2017a). China's three biggest unicorns – Xiaomi, Didi, and Meituan-Dianping – were all valued at more than US$10 billion and all are headquartered there. Despite having limited revenue sales outside of China and to some extent Asia, China's unicorns do remarkably well when compared to worldwide unicorns. For instance, *Fortune Magazine*'s 2016 Unicorn List ranks Xiaomi in second place behind Uber (Fortune Magazine 2016). In fact, three of the top ten unicorns on the list are Chinese companies.[3]

In an effort to increase the number of high-tech successes, China announced in 2016 that Zhongguancun will receive US$1.5 billion to update existing infrastructure; eliminate or relocate existing retail stores along "Electronics Street" to make way for building new, innovation centers; and to generate increased international attention to the park's high-tech industry (Business Standard 2016).

CHINA'S UNICORNS:
HOW INNOVATIVE ARE THEY?

Xiaomi

Xiaomi has been lauded as one of the most successful companies to originate from Zhongguancun. Founded in 2010 by Lei Jun, Forbes Asia's 2014 "Businessman of the Year," Xiaomi is most notably known as a provider of cheap smartphones. The company's flagship smartphone, known as the Mi, retails for around US$250–350 while its more afford-able version, the Redmi, is below US$150, and its top-of-the-line device, the MIX 2, is priced slightly over US$500. Comparatively, both Apple and Samsung price their top-of-the-line devices at around US$1000. One of the contributing factors to Xiaomi's ability to offer such competitively priced products is that it predominantly relies on online sales and has very few brick-and-mortar stores, cutting down on costs. In its heyday in 2014, it was valued at US$45 billion and was the third largest smartphone manufacturer in the world, behind only Samsung and Apple (Tilley 2016). Lei Jun credits part of the company's success to being headquartered at Zhongguancun:

> both [talent and capital] are readily available [at Zhongguancun]. I think it's fair to say that Xiaomi would not have been able to reach our current level of success without the outstanding financial support and incubator capability available in Haidian Park. To a large extent, our success was possible because we grew up in Zhongguancun and set down our roots in Haidian Park. (PRNewswire 2015)

Despite its success, Xiaomi is still not a common household name outside of China or Asia. Moreover, in recent years, it has been replaced by Huawei as the top smartphone seller in China. In an effort to combat decreasing sales, Xiaomi has announced that it plans to establish 1,000 brick-and-mortar stores in China and another 1,000 abroad by 2019 (Wang 2017b). In addition, Xiaomi is looking to expand to other markets with a focus on developing countries; it has done well thus far in expanding its presence in India. Xiaomi captured 23.5 percent of India's mobile market, along with Samsung, in the third quarter of 2017 and has become the country's fastest growing smartphone brand (Indo Asian News Service 2017). Xiaomi has yet to make its entrance into the US or Western Europe. Limited Xiaomi devices (e.g., home security cameras, Bluetooth headsets), not including smartphones, are offered to US consumers, but the company does not have any concrete plans to expand into the US (Doud 2017; Lai 2017). One of the main obstacles Xiaomi faces in trying to enter Western markets is the issue of patents, as it did in India where Ericsson sued Xiaomi for patent infringement on its wireless technology. The bulk of Xiaomi's patents lie within China, and the company faces many legal issues in bringing its products to countries with highly secured patents from competing companies such as Apple and Samsung. To prevent potential legal battles over intellectual property as it enters the US market, Xiaomi has been adding heavily to its patent portfolio, and has acquired over 1,500 patents from Microsoft along with patents from numerous other companies (Tilley 2016; Doud 2017).

Baidu

Baidu, China's biggest Internet search engine, was co-founded in 2000 by Robin Li and Eric Xu. Both studied and worked in the US prior to returning to China and co-founding Baidu. Xu received his PhD in Biology from Texas A&M University and worked as a post-doctoral researcher at the University of California, Berkeley from 1994 to 1996. Li, Baidu's current CEO, received his Masters in Computer Science from the State University of New York (SUNY) at Buffalo. After graduating, Li worked at a subsidiary of Dow Jones where he was involved with designing the financial information system for the online version of *The Wall Street Journal*. During this time, he received a US patent on a site-scoring algorithm for search engine page ranking (Baidu 2017), which was later used for the Baidu search engine. Li later joined Infoseek, a Silicon Valley search engine startup, prior to returning to China.

Today, Baidu has more than 70 percent of the market share in China and is the second largest independent search engine in the world (Baidu 2017). In 2017, the company expanded its mapping services to offer global coverage by partnering with mapping company Here (Cuthbertson 2017). Predominantly viewed as a service for Chinese tourists traveling abroad, Baidu's expansion of its mapping services puts the company in direct competition with other online mapping services such as Google Maps. The company has stated that it is aiming to have half of its map users coming from outside of China by 2020 (Cuthbertson 2017).

Like Google, Uber, Tesla, and other advanced technology-based companies, Baidu has been investing heavily in artificial intelligence and autonomous driving. It recently opened a Seattle research lab to focus on these efforts (Levy 2017). In September 2017, Baidu announced a 10 billion yuan (US$1.51 billion) "Apollo Fund" that will provide funding for 100 autonomous driving projects over the next three years (Cadell 2017). The company began its autonomous driving R&D in 2014 and its Autonomous Driving Unit (ADU) has headquarters in both Beijing and Silicon Valley (Baidu 2018). Baidu is working with an ambitious timeline, aiming to "commercializ[e] autonomous driving by 2018 and achiev[e] mass production by 2020" (Baidu 2018).

Tencent

Tencent Holdings Limited, a mobile and Internet value-added services provider, was founded in November 1998 and located at the Shenzhen High-Tech Park.[4] Tencent was co-founded by Ma Huateng, three of his college classmates, and a friend. Their first product was cloning an Israeli-made instant messaging service and adapting it for the Chinese market. Though Ma remains the CEO of Tencent, Martin Lau, who was educated in the US and worked for Goldman Sachs before joining Tencent in 2005, is Tencent's President and lead strategist.

Tencent's leading platforms are QQ, an instant messaging software service, and WeChat (also known as Weixin), a social media networking application. As of March 2017, QQ had 861 million monthly active user (MAU) accounts and WeChat had 938 million MAUs (Tencent 2017). WeChat began in October 2010 and is China's largest social network platform. It is difficult, however, to know how it would fare if competitors such as Facebook, Twitter, and WhatsApp were not blocked in mainland China. By comparison, Facebook cleared 1.9 billion

MAUs in May 2017 (Fiegerman 2017) and Twitter had 328 million MAUs in the first quarter of 2017 (Sparks 2017).

WeChat's reach, however, has not extended much beyond China's borders. There were only about 70 million WeChat users (~7.5 percent) outside of China in 2016 (Beaver 2016). Accessing foreign markets will be necessary if WeChat expects to grow. It has been estimated that approximately 93 percent of individuals in Tier 1 cities (e.g., Beijing, Shanghai, Shenzhen) are already registered WeChat users (Beaver 2016). WeChat's foray into the international market has not been very successful. Its attempted expansion into India, Indonesia, and Brazil have largely been considered failures (Parker 2017).

There are several reasons why WeChat has thus far failed at expanding internationally. By the time WeChat tried to expand to other countries, it was late in the game: WhatsApp had already taken root in the United States, while Blackberry Messenger was dominating in Indonesia (Custer 2016; Parker 2017; Stone and Chen 2017). Breaking into a new market is extremely difficult when potential customers are already using a different platform. For example, even if the Chinese government lifted its ban on WhatsApp, it would be very difficult for the app provider to gain a foothold in China since most users are already using WeChat. Another challenge (not only for WeChat, but also for Chinese companies more generally) is that most of its products and services are designed only for the Chinese market, which itself is fairly closed, and as a result have been unable to competitively meet the needs of the larger, international market (Custer 2016; Parker 2017; Stone and Chen 2017).

Didi chuxing (formerly Didi Kuaidi)

Didi is a transportation hail-riding service provider and often regarded as China's answer to Uber. Probably the least well-known of the four profiled companies outside of China, Didi Chuxing is headquartered at Zhongguancun and arose from a 2015 merger between two competing hail-riding services – Didi Dache and Kuaidi Dache. Didi Dache was founded in 2012 by Cheng Wei, a former Alibaba employee, who first got the idea for Didi from Hailo, a UK ride-hailing startup. He thought that the Hailo model could be emulated successfully in China. Kuaidi Dache was a successful competitor that was backed by Alibaba. The two startups merged in 2015 when Uber started moving into China. Didi Dache and Kuaidi Dache realized that one, or both, of the companies might lose out to Uber but that they might succeed together. What followed was a fierce competition between Uber China and Didi Chuxing for dominance in the Chinese market, with both companies, at some point, spending billions of yuan in unprofitable subsidies to attract both drivers and passengers (Stone and Chen 2016). Competition between Uber China and Didi Chuxing intensified in 2016, when Apple invested US$1 billion in Didi while Uber raised US$3.5 billion from Saudi Arabia's Public Investment Fund. In a situation reminiscent of the Didi Dache and Kuaidi Dache merger, Uber China and Didi Kuaidi concluded that it would be more profitable for both companies to come together rather than continue wasting money in competition with each other. In August 2016, Uber China and Didi Chuxing signed a merger deal in which Uber China officially left the Chinese market but would own 20 percent of the merged Chinese company and Didi would make a US$1 billion investment in Uber global (Mozur and Isaac 2016; Weinberger 2016).[5]

From imitation to innovation?

Baidu, Tencent, Xiaomi, and Didi are considered some of China's most successful, innovative companies. However, at the end of the day, Xiaomi did not invent the smartphone, nor did it popularize it to the international market; Baidu's search algorithm was patented by Li when he was working in the US; Tencent's WeChat does not have an extended user base outside of China; and the idea for Didi was copied from a foreign startup. In some aspects, a main contributing factor to the success of these companies – with the exception of Didi, which found a way to come together with its competitor – is that they face limited competition in China. WeChat's large success in China is due, in some part, to the fact that almost all foreign social media platforms such as Facebook, Twitter, Instagram, WhatsApp, and Snapchat are banned. Similarly, Baidu's success is unhampered by the fact that Google left China in 2010 because of disagreements with the Chinese government regarding censorship and accusations that the Chinese government had made a sophisticated and coordinated attack on Google's corporate network. These companies are very successful within the Chinese market and have enjoyed some success within Asia. However, none have much reach outside of China or Asia, and have limited name recognition among Western consumers.

How innovative are these companies? Their success lay in re-innovating popularized products or technologies for the Chinese market; they did not, however, invent the underlying technology or its application. Chinese companies, successful as they may be in China, are still very much relying on second-tier innovations (Breznitz and Murphree 2011). Recently, some of these companies' innovative features have drawn the attention of their international competitors such as Facebook and Amazon who have started to learn their playbooks (Lucas 2018). Furthermore, they have benefited immensely from the fact that the Chinese government bans much of the foreign social media and Internet service providers from accessing the Chinese market.

NANOTECHNOLOGY: HAS CHINA SUCCEEDED IN ACHIEVING INDIGENOUS INNOVATION?

To assess the degree to which China has achieved its goal to become globally innovative, our own research[6] has examined one frontier technology as a focal point of this chapter: nanotechnology.[7] We focused on nanotechnology both because of its far-reaching potential as a platform technology that can reach across diverse industries, and because China's MLP identified nanotechnology as one of the initial four "science mega-programs" that were seen as key drivers of indigenous innovation. Nanotechnology provides an excellent case study of China's S&T efforts, since it is regarded as a leading-edge technology – one that was predicted early on to herald a technological revolution capable of solving many human problems, while generating enormous economic returns (Lieberman 2005: xi; Roco et al. 1999: iii).[8]

The list of nanotechnology's promised benefits is seemingly endless, and includes low-cost hybrid solar cells providing cheap energy; targeted, non-invasive, and thereby less toxic drug delivery, achieved by constructing nanoscale particles that migrate and bond with specific types of cancer cells, which are then selectively destroyed; "lab-on-a-chip," providing instant diagnosis of multiple diseases in remote field settings; ultra-high-speed computing, thanks to nanoscale data storage devices; and highly efficient nanoscale filtration at low cost, providing a solution for air pollution and water contamination.[9] It is estimated that as of 2016 some 10,000 firms globally were engaged in nano-related research, development, and manufacturing – of which more than half were from the United States (US NAS 2016: 12).[10] While it is difficult to predict with any certainty the future commercial value of nanotechnology, one recent report estimates that the global value of nano-enabled products, nano-intermediates, and nanomaterials will have reached US$4.4 trillion by 2018 (Lux 2014) (see figure 5.2).

Both because of its commercial potential, and because the US in 2000 launched a National Nanotechnology Initiative intended to win a global race for a predicted multi-trillion–dollar market of nano-enabled products, as we have previously noted, nanotechnology was included as a science mega-program in the MLP. Given its interdisciplinary nature and long-term commercial prospects, nanotechnology was unlikely to be part of the MLP's engineering mega-programs. In order to raise

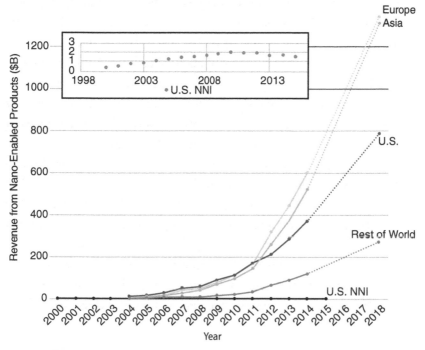

Figure 5.2 Revenue from Nano-Enabled Products: Selected Countries
Source: US NAS, 2016: 15. Reproduced with permission conveyed through Copyright Clearance Center, Inc.

nanotechnology's profile, it was necessary for its champion, CAS President Bai Chunli, to emphasize its basic science aspect while using the potential application bonanza as the attraction (Appelbaum et al. 2011b; 2012; Cao et al. 2013b). With its vast resources in foreign reserves and long tradition of state planning, China hoped to become a global player in nanotechnology. China is estimated to have invested upwards of US$1 billion during the 15-year period from 2000 to 2015, including 141 nano-related projects in such areas as nanomaterials, nanoequipment, nanoenergy, nanocatalysis, and nanomedicine, since nanotechnology was singled out in the MLP (Qiu 2016).

How successful has this effort been?

One indicator of China's success in basic research is the rapid growth of nanotech publications, one area in which China is a rising star. China's share of global publications has risen sharply during the past two decades (see figure 5.3). As China's nanotech S&T system grew in strength and numbers beginning around the year 2000, international collaboration briefly declined because China's scientific community turned inwards in search of collaborators – a trend that has reversed in recent years. One possible reason is that while ramping up its efforts at indigenous innovation, China increasingly partnered internationally in basic research – a trend that most likely contributed to a growing share of Chinese-co-authored English-language papers (by some estimates as much as 90 percent) (Appelbaum et al. 2011a; Mehta et al. 2012). China's output of SCI publications now exceeds that of the US (Qiu 2016), although their impact (as measured by citations) is considerably lower (Han and Appelbaum 2016). China also exceeds the US in the estimated number of nanotech researchers (US NAS 2016).

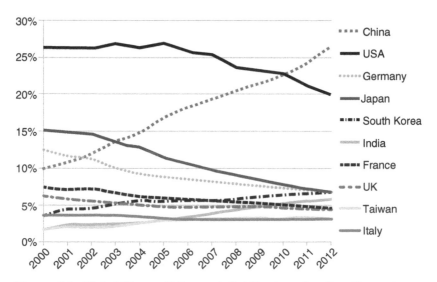

Figure 5.3 Global Share of Nanotech Publications: Leading Countries
Source: UCSB-CNS Analysis

China has made progress in terms of nanotech R&D, at least as indicated by patents. Drawing on our patent dataset (106,000 patent families based on EPO's PATSTAT), when compared with the top four countries in terms of global nanotechnology patent family counts, in 2013 China accounted for 27 percent worldwide; the US, 16 percent; South Korea, 13 percent; and Japan, 9 percent. China's share had nearly doubled since 2008. Nearly two-thirds of China's nanotech patents were from the academic sector (universities or the CAS), and all but one of the top five most frequent nanotechnology patent applicants were from academic institutions representing China's most elite universities.[11] Only a sixth of all patents were found to be corporate, with roughly another sixth from government-sponsored projects. This suggests that the large majority of nanotechnology patents in China remain closer to basic research than to development, with relatively few pertaining to marketable consumer products – evidence that China continues to face challenges in achieving its goal of transferring and translating academic research into viable products (Appelbaum et al. 2011a; Cao et al. 2013a; Parker and Appelbaum 2012). Moreover, the share of carbon nanotubes, surfaces, and substrates has become more prominent among China's top ten patent areas (see figure 5.4). These are primarily areas that are fairly low on the nanotechnology value chain, providing materials that are incorporated in the products of (non-Chinese) multinationals. Emerging as a world leader in carbon nanotubes and graphene, at a time when these are becoming low-cost commodities, is unlikely to result in the indigenous innovation that China seeks to achieve.

China's growth in nanotech publications and researchers obscures many basic challenges in its research environment, discussed below, leading one leading US-based Chinese nanoscientist to conclude that "while China has the largest number of patents and SCI publications in nanotechnology, only a tiny minority of them are truly innovative" (Yang Peidong, quoted in Qiu 2016: 150).

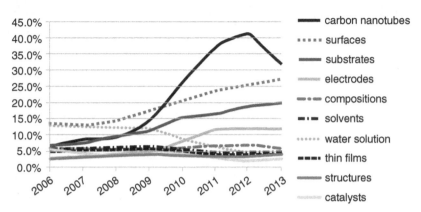

Figure 5.4 China's Top Ten Nanotech Patent Areas
Source: Appelbaum et al., 2016

The Chinese government has pursued several strategies to foster a better payoff between basic research in nanotechnology and eventual commercialization of nano-enabled products. All of these have come with some costs (Appelbaum et al. 2016). For example, the devolution of policy to the provincial and municipal levels has resulted in uncertainty, since rapid shifts in national priorities can (and often do) affect local funding. Direct government funding for R&D and commercialization, often through the Ministry of Science and Technology, greatly exceeds peer-reviewed competitive funding for basic research through the National Natural Science Foundation of China. This both hampers the basic research needed for innovative advances in nanotechnology, and often results in *guanxi* (personal) relationships guiding decisions. China's state-owned enterprises (SOEs), which are favored when it comes to public investment, tend to be bureaucratic, risk (and hence innovation) averse, and beholden to party connections.

Nanotechnology-related R&D remains spatially concentrated, resulting in limited technological spillovers that might contribute to the growth of regional R&D centers. Between 1986 and 2008, nanotechnology patents became concentrated on China's east coast, with the

Beijing and Shanghai regions becoming increasingly dominant; the greater Shanghai region (which includes Jiangsu and Zhejiang provinces) had surpassed the Beijing–Tianjin region by 2007. The geographic concentration of patents within regions is small (around 20 km), suggesting limited spillover effects. Patents originating in universities increased over the period, while those originating in industry stagnated, suggesting an absence of technology transfer from university to industry; academics typically lack the experience and resources to turn ideas into products (Motoyama et al. 2013). Bai Chunli, CAS President and widely considered the driving force behind Chinese nanotechnology, recently commented:

> The new government sees innovation as central to China's long-term development. But most researchers in China are keen to follow hot and trendy research areas. Few have the spirit of *shi nian mo yi jian* ("taking a decade to sharpen a sword") or are willing to pursue big research questions. For instance, China has invested heavily in graphene and carbon nanotubes after they were first developed in the West, but very few genuine innovations have come out of it. We really have to unleash our creativity rather than being content to follow the crowd. (Bai, quoted in Qiu 2016: 150)

In sum, while progress has been made, China has yet to become a nanotechnology innovator; its capacity for nanotech innovation remains behind that of the US and other advanced industrial economies. Although China's state policies (and funding) have had some success in advancing research, they have been less successful in bringing viable products to market. China's effort to commercialize nanotechnology has been much slower than anticipated by China's nanoscientists and leadership. Even though basic research is improving, it still fails to drive long-term innovation; as a result, products typically mirror the functionality of existing products, rather than representing breakthroughs. Many subfields of nanotechnology remain at an early stage, although this is a problem not unique to China, since advances in basic research outpace technological applications in all countries.[12] China, however, is faced with the additional challenges described above: to paraphrase a Chinese proverb, "research is high and the market is far away" (Cao et al. 2013b).

SUZHOU INDUSTRIAL PARK: SHOWCASE FOR NANOTECHNOLOGY?[13]

Suzhou Industrial Park (SIP) was created to directly foster commercialization by providing supportive infrastructure and venture capital (VC) funding (Cao et al. 2013b; Appelbaum et al. 2016). As an emerging technology, nanotechnology provides a case study of such efforts, in an area of basic science where all countries have entered at roughly the same time; none has (as yet) a first-mover advantage.

SIP, founded in 1994, is a county-level administrative jurisdiction covering some 111 square miles (SIPAC 2015). Its website unabashedly (if questionably) promotes it as "Creating a new Silicon Valley and Realizing Technological Leap Growth."

> Suzhou Industrial Park is the first among national development zones to implement the three action plans of industrial upgrading, service sector multiple growth, and technological leap growth, actively promoting the integration of using international capital and learning cutting-edge technologies so as to promote the industrial transformation and upgrading, and to switch from "made in Suzhou" to "created in Suzhou." (SIP 2009)

SIP is a development zone of the prefecture-level[14] city of Suzhou in Jiangsu province.[15] Suzhou, founded 2,500 years ago, is part of the Yangtze River Delta region; the ancient city has long been noted for its livability and is one of the richest cities in China (Van Winden et

Box 5. Smart Cities

In connection with the Internet of Things (IoT), China has been focusing on the development of (and indeed transition to) "smart cities" that will use advanced information technologies such as IoT, cloud computing, and big data analytics to improve infrastructure, governance, and public goods and services. By 2013, China had already approved 193 smart city pilot projects (Li et al. 2015), and it was estimated that the number of smart cities in China would exceed 500 by the end of 2017 (*China Daily* 2017c). Comparatively, in a 2016 survey, 54 US cities indicated that they have 335 implemented and 459 planned smart city projects through the end of 2017 focusing primarily on governance, mobility and transport, and physical infrastructure (United States Conference of Mayors 2017).

A prime example of China's efforts in building smart cities is the city of Wuxi. Designated as China's sole national demonstration zone for IoT in 2009, and selected as one of five core cities in the world to participate in the Institute of Electrical and Electronics Engineers' (IEEE) Smart City Initiative in 2014, Wuxi is aiming to become a safer, more efficient, and more responsive city through IoT and other advanced information technologies by 2020 (Wang 2016; IEEE Smart Cities 2017). Home to more than 2,000 IoT-based companies, Wuxi's IoT market is expanding rapidly and in 2016, IoT accounted for 210 billion yuan (US$31.6 billion) of the city's total annual revenue (Hu and Lu 2017). China has positioned itself to become a leader in both IoT and smart cities. However, unlike many of its Western counterparts, China has traditionally placed much lower emphasis on protecting user privacy and security. Unless China is able to address these concerns, it may have difficulty in marketing any future advances and innovations in IoT to Western consumers.

al. 2014). Situated on several lakes and famous for its canals and gardens, Suzhou is variously described in SIP's promotional materials as "paradise on earth" (SIPAC 2012), or "the beautiful ancient garden city ... known as the Venice of the Orient" (SNC 2011: 2). The city is 50 miles (and 20 minutes by high-speed train) west of Shanghai, with ready access to airports in Shanghai and Wuxi; it is also served by the high-speed rail service on the Beijing–Shanghai line.

SIP was originally conceived as a new industrial city, inspired by Singapore's experience in successful planned urban development. In 1994 China's then-Vice-Premier Li Lanqing and Singapore's then-Prime Minister Lee Kuan Yew met in Suzhou, signing an agreement that created the China–Singapore Suzhou Industrial Park (CSSIP) to the east of the historic city of Suzhou. The Singapore consortium originally had a 65 percent share in the park, while a Chinese consortium held the remaining 35 percent. The China–Singapore partnership was to be "a model of mutually beneficial cooperation" (SIP 2014): Singapore would show China how to effectively court international investors, while gaining an entrée into China's growing role as the world's leading manufacturer. Singapore would provide a model for China's nascent business environment: stability, highly competent and dependable management, and superior infrastructure (Pereira 2003).

It did not turn out that way, however. CSSIP lost money during its early years because Suzhou was also investing heavily in a competing industrial park just to the west of the city. The Suzhou New District (SND) had preceded CCSIP by several years, and was able to attract global firms such as Motorola, Sony Chemicals, and Philips Electronics even as CCSIP sought to expand. By 2001, in what Singaporeans viewed as a scandal, the Singapore consortium reduced its share to 35 percent (Dolven 1999).[16] With China now the senior partner, SIP took off, leading to speculation that the Chinese had deliberately favored SND over CCSIP in an effort to wrest control (SIP interview 19 April 2012; for an extensive discussion, see Zhao and Farole 2011).

SIP today encompasses six zones: the new, ultra-modern Central Business District for the city of Suzhou around the man-made Jinji Lake; the Dushu Lake Science Education Innovation Park; the Yangcheng Lake Eco-Tourism Resort; a high-tech Industrial Zone; an ecological science hub; and an Integrated Free Trade Zone (SIPAC 2015). Suzhou has also created a number of national-level development zones, including the Suzhou High-Tech District,[17] the Zhangjiagang Duty-Free District, the Kunshan Economic and Technological Development Zone, and the Suzhou Taihu Recreation and Holiday Zone (Wei et al. 2009: 416).[18] In 2002 Suzhou also established the Dushu Lake Higher Education Town within the Science and Technology Innovation District, home to 18 of Suzhou's nearly two dozen colleges and universities,[19] including branches of several universities from the UK (Liverpool, Warwick, Limerick), the National University of Singapore, the Hong Kong University of Science and Technology, and Dayton University in Ohio (Wei et al. 2009: 422). The entire park is intended to function as a model garden city, integrating all aspects of a vibrant urban life: higher education, cultural amenities, and outdoor recreation with a wide range of support for manufacturing and innovation. As such, its ambition is to provide a sufficiently attractive living environment for foreigners and Chinese alike, effectively competing with Shanghai, Beijing, and other major cities.

In some aspects SIP has clearly been successful, offering its tenants an industrial ecology designed to foster innovation. The industrial park provides expedited project approval; issues passports and requests visas; and provides a provident fund (based on personal account deposits) covering medical expenses, housing, retirement, and social assistance (Wei et al. 2009: 417). It provides a "one-stop, full package" range of services, including a Software Park; a Patent Exchange Navigation Center (enabling tenants to identify, buy, and sell existing patents); an Incubator Center; favorable tax treatment, including tax holidays; export processing zones (EPZs) that include expedited logistics and exemptions from various import and export duties and taxes; a highly efficient customs office and tax bureau, resulting in fast customs clearance; and a Bonded Logistics Center (Wei et al. 2009: 417–19). Several years ago, its promotional materials boasted having 160 R&D institutions, 85 venture capital entities, and more than 100 national-level breakthrough projects (Suzhou Industrial Park promotional video, viewed May 2012). As a result of these services, SIP now claims to be home to thousands of national and multinational companies, including some 90 Fortune 500 companies such as Nokia, Fujitsu, Mitsubishi, Daimler Chrysler, BP, ZF, 3M, Samsung, Siemens, Johnson & Johnson, Philips, AMD, Bosch, and Eli Lilly. It also claims to generate 15 percent of GDP for the local economy, having attracted US$26.7 billion in foreign investment for 5,200 programs (Xinhua 2015; HKTDC 2015; Suzhou Industrial Park promotional video, viewed May 2012).

As with everything that relates to China's development policies and trajectory, there is disagreement in the scholarly literature over whether or not SIP has been successful in achieving its vision as a high-tech park. One recent study, conducted for the Ministry of Science of the German state of Baden-Wuerttemberg,[20] concluded that the outlook for SIP was extremely good: "The numbers alone speak volumes: every second day, a foreign company commences business within the boundaries of SIP" (Lanza et al. 2015: 116). While generally bullish on SIP (and China in general), the study did note some challenges, including increased labor costs in China, anticipated import tax increase on high-tech components, and high levels of pollution (which may discourage foreign entrepreneurs from relocating to China).

Another recent study (Jin et al. 2015) concluded that SIP provides an example of what has been termed "knowledge-intensive entrepreneurship" (McKelvey and Lassen 2013), in which a firm's success reflects the interplay of knowledge inputs, founders' characteristics, financing, and favorable societal influences (such as high-tech industrial parks). The founders and leading managers of four high-tech SIP firms (all involved with nanotechnology, all successful) interviewed by us[21] all shared some common characteristics: their founders or co-founders had conducted university research or worked in companies engaged in nanotech, and they all had close links between science and industry from the start.[22] SIP was seen as playing a key role in supporting knowledge-intensive entrepreneurship by helping the firms to identify sources of research funding and venture capital, as well as providing a venture capital fund of its own.

Importantly, SIP's overall environment was seen as crucial for the four firms' success in innovation. SIP provided land and office space; all four companies were given space in Nanopolis (described below), which enhanced their networks with potential suppliers and clients and facilitated cooperation that resulted in technological advances. SIP provided numerous amenities: the benefits of an eco-friendly city; education (from kindergarten through university); a variety of services aimed at startups; low-rent apartments for firm employees; salary subsidies for those with advanced degrees; and industrial production facilities. SIP also has campus branches from many universities, providing both an intellectual (and educational) environment for park residents, as well as support for research and development. The firms also benefited from the presence of the Suzhou Institute of Nanotech and Nanobionics (SINANO), a CAS institute "dedicated to the basic, strategic and forward-looking researches in nano materials and devices, nano biomedicine, nano-bionics, nano safety and other fields," and offering "nano analysis and test platform, nano processing platform and other open facilities" (Jin et al. 2015). The study concluded that firms were attracted by the

> eco-system of SIP and moved to there for the industrialization of nano-technologies, ... entrepreneurial ventures in nano can find important partners for network relationships – such as institutes, suppliers, and even customers – in the park, as well as access to scarce resources such

as new office space, and access to special policy measures to stimulate business based upon advanced knowledge. (Jin et al. 2015: 161–2)

On the other hand, in the classical Marshallian formulation (Marshall 1920 [1890]), a successful park (or "industrial district," in the original formulation) offers "flexible specialization and agglomeration economies with a synergistic combination of dense local networks, local innovation, and small firm formation and clustering" that contribute to "regional growth and competitiveness" (Wei et al. 2009: 411). This classical notion has since been modified to take globalization into account. "Neo-Marshallian districts" are seen as key industrial centers in the global economy that simultaneously provide spillover agglomeration effects for the local economy (Amin and Thrift 1992) while "satellite districts" are dominated by foreign firms with little local spillover (Markusen 1996).

In one study of SIP – based on research that began in 1998 and concluded in 2007 – Dennis Wei, Yuqi Lu, and Wen Chen's "overall assessment of the Suzhou pathway to industrialization and regional development is positive" (Wei et al. 2009: 424). But, in their view:

> Suzhou is not yet a neo-Marshallian district; it is still a satellite district dominated by TNCs and external organizations. The most serious challenges to Suzhou's development are the lack of top-ranked research institutions, which are required for research and development (R&D) and the development of high-technology industries, and the focus of the city on manufacturing functioning as a global manufacturing floor. It is also a rapidly expanding city fragmented among the old city district, Suzhou New District (SND), and SIP. In addition, Suzhou is located in the fragmented Yangtze Delta and experiences intense competition with Shanghai. (Wei et al. 2009: 424)

Others have also examined Suzhou and its industrial roots (the city is home to at least four industrial parks, of which SIP is the largest and most impactful). In their case study of SIP, Van Winden and colleagues note that SIP "stands out as the largest and most influential industry cluster in the Suzhou region ... as one of the leading parks in China" (Van Winden et al. 2014: 132). The post-2007 period is crucial, since SIP launched its industrial upgrading campaign in 2005, focusing on advanced manufacturing and high-technology sectors. Since that time, and consistent with China's MLP and 11th and 12th Five-Year Plans, SIP has aggressively pursued a course of high-tech innovation.[23] Although SIP began as a fairly conventional industrial district in the hope of attracting foreign and domestic manufacturing, in recent years it has increasingly emphasized high-tech innovation and entrepreneurship. It is currently home to some 24,000 Chinese firms, 3,800 foreign firms, and 130 SOEs (Lanza et al. 2015). According to Liu Hua, who heads SIP's Bureau of Services, more than 2,000 firms now provide services to park clients; "more and more companies are now turning to areas like big data, cloud computing, mobile internet, and the Internet of things, and they are staying away from low value-adding activities such as labor intensive outsourcing work" (Shen 2015). SIP also claims to have attracted more returnees (118) under China's Thousand Talents Program than any other industrial park. In 2010 China's Ministry of Commerce ranked it second in performance (behind Tianjin's Economic-Technological Development Area) among 90 industrial parks (Shira 2011). The Chinese government also has designated SIP as an eco-industrial park, and as a "pilot zone" for IPR protection and preferential policies for high-tech services (Zhao 2012; CSSD 2014; Lanza et al. 2015). A World Bank study described SIP as having "built a reputation as one of the most business-friendly, residential-friendly, and environment-friendly industrial parks in China" (Zhao and Farole 2011: 103).

In October 2015, China's State Council approved a pilot zone program proposed by the Ministry of Commerce and the Jiangsu provincial government that would turn SIP into "a world-class high-tech industrial cluster and enhance international cooperation." The plan called on SIP to attract more high-tech industries, including high-end factories of multinational corporations; encourage resident firms to emphasize R&D and marketing; urge Chinese patent holders to partner with multinationals in an effort to more effectively commercialize their ideas; and enable SIP's Chinese firms to purchase advanced technology and equipment. SIP industries specifically singled out for additional investment included biopharmaceuticals, information technology, cloud computing, and – significantly – nanotechnology (China State Council 2015).

Nanotechnology in SIP

As we previously have noted, we focus on nanotechnology as an entrée into high-tech development because it is believed to be a transformative technology at the forefront of high-tech innovation, and is one of four "science mega-projects" called for in China's MLP and recent five-year plans. Since it is seen as having the potential to leapfrog China into a high-tech future, it provides a useful case study for assessing China's innovation potential. While China's success in IT was largely the result of its ability to build on already established technologies (Breznitz and Murphree 2011), this is not the case with nanotechnology. China launched its national effort at roughly the same time as the US, Europe, Japan, and other advanced economies. In sum, by focusing on nanotechnology, we hope to shed some light on whether China's substantial public investment in SIP, with its vast pool of scientists, engineers, and entrepreneurs, has resulted in the nanotech breakthroughs envisioned by China's leadership.

SIP launched a major effort to promote nanotechnology research and development in 2011. Among all high-tech parks in China, SIP is the only one with a designated specialization in nanotechnology. It was designated by MOST as the "China International Nanotech Innovation Cluster," with 10 billion yuan (US$1.6 billion) to be spent over a five-year period in support of its "Nanopolis Suzhou" Initiative (as previously noted, a play on Singapore's successful Biopolis) – an entire geographic sector intended to provide shared resources and a collaborative environment for young nanotech firms. Nanopolis, envisioned as the "nanotech commercialization hub in China," is intended to provide "a complete ecosystem support for the growth of nanotechnology and its enabling industries" (Suzhou Industrial Park 2011: 2). It claims to be the world's largest hub of nanotech innovation and commercialization, with a floorage of 100 acres and a planned construction area of 1.5 million square meters. Phase I was completed in 2012, and construction on phase II had begun; it was predicted to cost 6 billion yuan (US$950 million) to construct and cover 250 acres (0.4 square miles). By August 2014, 80 nanotech-related enterprises and service agencies had settled there, with two major business clusters – nano new materials and microelectromechanical systems (MEMS) – starting to take shape. With the establishment of China–Finland Nano Innovation Center, Holland High-Tech China Center, Czech Tech China Center, among others, the international nanotechnology industry appear to be using Nanopolis as the launching base for its technology transfer and commercialization activities (Nanopolis 2015).

SIP describes its R&D support facilities as "world class ... open platforms" (Suzhou Industrial Park 2011: i), including (from its self-description) (SIP Nanopolis 2018):

+ **Nano Fabrication Platform:** the only "24/7" nano manufacturing platform in China, open to startups, R&D institutes, and universities.
+ **Nano Characterization and Testing Platform:** open, shared platform with world-class facilities and services available to users from research institutes and companies.
+ **Printed Electronics Research Center:** the first and only printed electronics research center in China, established as a national flagship.

+ **Functional Nano Carbon Materials Group:** comprehensive facilities for scalable CNT array synthesis, high-end, electronic device fabrication and testing.
+ **Pilot Production Platform for L-ion Battery:** covering materials, manufacturing, management and testing.
+ **High-efficiency Photovoltaic (PV) Center:** comprehensive facilities for high-efficiency PV fabrication, integration, and characterization.

According to its director, Nanopolis is projected to have 300 tenants and as many as 30,000 employees by 2019 (Shen 2014). Nanopolis aims to turn itself into a nanotechnology industrial ecosystem formed around several essential elements including:

+ **International platform.** Nanopolis has partnered with over 20 countries and regions across the world, including Finland, Holland, Czech Republic, Germany, Korea, the US, Japan, and France.
+ **Industrial investment.** Suzhou Nano Venture Capital Co., Ltd., China's first fund focusing on the investment and incubation of nanotech startup projects, has contracted 10 nanotech projects worth 100 million yuan (US$15.1 million) with renowned Chinese venture capitalists.
+ **Technical Industry Platforms:** Nanopolis provides four state-of-the-art public platforms – fundamental research, applications research, engineering R&D, and public technical service through the Gallium Nitride (GaN) Materials and Devices Research Institute, R&D Center for Nano-Lithium-Ion Battery Materials, and SIP Nano Safety Assessment Center.
+ **Intellectual Property generation and protection.** SIP Nanotechnology Patent Operation Center pools resources from China's State Intellectual Property Office to implement the micro-navigation strategy and to provide guides to enterprises in respect to patent layout and technological innovations and help them gain a competitive edge in the market.

+ **Nanotech Standardization:** With the support of Suzhou Demonstration Area for National Nanotechnology Commercialization Standardization, the SIP Nanotechnology Standardization Promotion Center is founded to pool all related domestic resources and push for the establishment of a national alliance for nanotechnology standardization and boost cooperation between nanotech enterprises and standardization institutions with the purpose of promoting the technological competence of enterprises to the market.

+ **Industrial Alliances:** Suzhou Alliance for Nanotechnology Applications Industry, Suzhou Nano Touch Industry Alliance, and SIP Laser Industry Innovation Alliance Association have been established with the joint effort of industrial leaders, upstream/downstream businesses along the industrial chain and the industry associations to advance R&D, applications docking, and cooperation in nanotechnology and nano products. The alliances have members from the Yangtze Delta area, East China, Middle China, and other parts of China.

+ **Product Promotion:** Nanopolis helps companies to promote the latest nano products through different media platforms, such as nanotech showrooms, nano magazine, website, microblog, and WeChat, as well as offering professional, systematic, and effective publicity and promotion services by combining public media and different exhibitions to help businesses with brand publicity, product exhibitions, marketing and promotion, technological cooperation, and sales channel expansion.

+ **Industrial Activities:** Nanopolis aims to establish diversified platforms for industrial exchanges so as to promote market applications of nanotech achievements through holding Go Go Nano, workshops, Nanotech Investors' Club, international communication activities, international project matchmaking, up/downstream matchmaking, project roadshows, and product promotions.

+ **Fund Application:** Nanopolis has set up professional teams to provide timely, considerate assistance to businesses in applying for funding schemes for key technological projects, leading technological talent projects, etc., to promote the R&D, product design, and market expansion of companies.
+ **Talent Training:** Nanopolis has excellent educational resources such as China Semiconductor Industry Association (CSIA), Interuniversity Microelectronics Centre of Belgium, Tsinghua University, and Chinese Academy for Sciences, etc., to give employees easy access to top-level training and education, and to improve their knowledge and skills to inject fresh energy into company development and technological innovations.

Nanopolis claims to provide comprehensive, integrated services for all aspects of nanotechnology R&D and commercialization, including incubators, pilot and mass production, corporate headquarters, and conference and exhibition facilities. Key focal areas include nano new materials, micro- and nano-manufacturing technologies (e.g., printed electronics, MEMS and NEMS, nanolithography, instrumentation), energy and green technologies (batteries, photovoltaics and lighting, gas/water purification, green technologies) and nano medicine (targeted drug delivery, imaging and diagnostics). Nanopolis also sponsors the annual Chinano conference and exposition in Suzhou, self-described as "the premier nanotech business event in China" (Chinano 2018). Chinano attracts thousands of exhibitors, presenters, academics, and business people, featuring the latest developments in Chinese nano-technology efforts at commercialization.

Nanopolis is the largest and most ambitious nanotechnology annex within the park, but it is not the only one. Close by to Nanopolis is bioBay, which was established in 2005 with an initial investment of US$68 million to "attract global companies with a focus on the develop-ment of bio- and nanotechnology" (SIP BioBay 2008). In fact, Nanopolis is a spin-off of bioBay. Located in Dushu Lake Higher Education Zone, bioBay offers an attractive campus-like environment, designed to attract scientists and engineers engaged in basic research, as well as firms that are commercializing biological products. The campus includes buildings dedicated to biotech incubation, nanotechnology, and the specific needs of bioBay tenants. It offers the usual blend of tax incentives, state-of-the-art equipment, legal and regulatory advising, talent recruitment and training, and a variety of government subsidies for promising firms (SIP BioBay 2008). bioBay claims to house nearly 300 companies, about a third of which are engaged in developing diagnostic medical devices, and about another quarter in new drug discovery.[24] Through such efforts, "Suzhou intends to attract over 200 nanotech companies from all over the world and 10,000 nanotech experts within the next 5 years to make Suzhou the most global and innovative nanotech hub in China by 2015" (Suzhou Industrial Park 2011: 2). Whether such goal is in fact attainable remains an open question.[25]

How successful are SIP's efforts to foster nanotech breakthroughs?

With a reported 10 percent annual growth rate, SIP is frequently touted as a desirable location for China's burgeoning group of high-tech entrepreneurs who are eager to co-locate side by side with foreign and domestic companies, and who are seeking encouragement to do applied, market-driven research. While SIP's early investments can be seen as staking a claim to China's hoped-for innovation economy, our interviews with researchers, entrepreneurs, and government officials[26] found that SIP's drivers for nanotechnology development are at least in part offset by a number of significant shortcomings.

Drivers for success may include such things as SIP's incentives to attract high-tech firms, as well as the potential regional economic advantage of the park, including its location in a wealthy province with high quality of life; and its easy connections to the world-class city of

Shanghai, which in turn links the area to the larger global innovation system. Through a series of incentives such as 50 percent matching funds for awarded research support offered by SIP, and various other support mechanisms such as incubation services, discount land acquisition, financial rewards for patents (individuals reportedly can receive 2000 yuan (US$300) for submitting a patent application alone and up to 5000 yuan (US$750) if the patent application is granted) and tax incentives, among others, SIP is an attractive landing point for overseas Chinese returning to mainland China after years spent abroad in the US, Australia, or Europe. Returning to China comes with benefits as well. Startup packages for researchers trained internationally are extremely competitive and at many universities and research institutes inside the park and elsewhere in China, promotions are contingent upon having significant foreign research experience. As one interviewee explained, "If you want to work with companies, and want to be encouraged to set up companies, to do application-oriented research, you'll be fine in Suzhou" (SIP interview, April 25, 2012).

Jiangsu province – which houses Suzhou – is an affluent part of the country which has, in recent years, invested heavily in education and in attracting high-tech industries, further making it a desirable destination for foreign-trained returnees seeking world-class amenities for their families. This point is echoed by several people with whom we spoke, but captured by one in particular, who commented that in his opinion "Jiangsu and Suzhou especially are very good at attracting and promoting small businesses because they have the best policies in all of China" (SIP interview, April 25, 2012). This same respondent recalled an early visit to Suzhou when he was first being courted to return to China – specifically Suzhou and SIP. He explained that he was "very impressed by my first visit to SIP in 2007 ... I didn't think a place like this could exist in China," he said noting that the world-class amenities that would be available to him – professionally and to his family personally – were encouraging. The reported rise in attractiveness of Suzhou – and SIP – to returnees may be at least in part attributable to China's focus on the development of its S&T ecosystem. Ultimately, while many interviewees commented on their personal preference and desire to stay in the US, several interviewees noted a lack of US employment opportunities commensurate with their experience as influencing their decision to come back to China.

Regardless of the hope and prosperity that Suzhou claims to provide, several interviewees were quick to temper their optimism, noting that "Silicon Valley will still be number one in high-tech innovation and entrepreneurship for the next 30 to 50 years but [during that time], Suzhou will be number one in China" (SIP interview, April 22, 2012). Looking forward, SIP's current attractiveness may suffer, of course, as a result of the economic recovery in Europe, the US, and Japan (Cao et al. 2013b).

Significant funding for the development of research institutes located within SIP come from the park itself as opposed to exclusively from the CAS or other government sources. It is typical for funding to be from a variety of sources, however, often including the CAS, Jiangsu province, the Suzhou city government, and SIP. This is often in addition to private sources of funding as well. Like private sources of funding (including venture capital as well as private/public partnerships), local and provincial funding typically come with an expectation of an economic development return on investment (SIP interview, April 25, 2012). In addition to expectations of economic development, local financial support for research institutes and startup companies comes with requirements to hire local staff as well as, increasingly, to file patents and transfer technology (SIP interview, April 22, 2012). It is also increasingly common for government funding to require a business plan from applicants (Chinano Interview, September 14, 2012).

Funds from venture capital groups can often surpass those from government for individual PIs. Such support however, does require loan repayment, as is typical of private sector support (SIP interview,

April 25, 2012). Adding to Suzhou's appeal as an innovation center is the fact that a considerable percentage of China's VC funding is in Suzhou or its surrounding area. One interviewee believed that this fact is one of many helping to propel Suzhou into prominence as what he predicted would become another Silicon Valley (SIP interview, April 22, 2012). Such venture capital – which is mainly from the government – tends to have a longer period for expected returns, which can encourage more innovative efforts that may require more time to bear fruit (Shanghai-based VC firm interview, April 20, 2012). True venture capital support – not backed by government – currently remains limited to a relatively small number of firms within SIP and China more broadly. Private venture capital companies that choose to operate within China are also plagued by the lack of an established, consistent legal framework for protecting investors and entrepreneurs (Batjargal and Liu 2004; Puffer et al. 2010; Zimmerman 2015). This weak regulatory environment leads private venture capital firms in China to adopt a more conservative investment strategy than venture capital companies operating in established markets (Zacharakis et al. 2007). With the government serving as quasi-VCs, one could expect SIP to emerge as a genuine and robust high-tech park. VC, however, currently remains limited to a relatively small number of firms. To give a sense as to how competitive China's VC support has become in recent years, a representative of one VC firm with whom we spoke commented that of the 300 to 400 business plans they receive each year, only three or four will be supported. The decisions are based on the perceived innovativeness of the company, the work experience of its leadership, as well as the market potential for the product in development (Shanghai-based VC firm interview, April 20, 2012).

Despite an increase in private sources of funding, the government is still the primary backer of new technological development, supporting as much as 90 percent of a startup company's costs (SIP interview, April 25, 2012). In addition, the government is also able to provide important tax breaks for low- or no-cost land acquisitions to young companies. It is these sorts of in-kind support mechanisms that are proving beneficial to SIP. When phase II Nanopolis construction is complete at SIP, the amount of space dedicated to nanotechnology companies and labs will double, with Nanopolis serving as the park's nanotech cornerstone. Nanotechnology companies and labs located inside such areas as Nanopolis and bioBay are under increasing pressure to perform, as measured by their funding levels, numbers of publications and patents, and the size of their talent pools (SIP interview, April 25, 2012). This last indicator is seen as one of the central drawbacks to being located in Suzhou, and unaffiliated with the CAS in a significant way. That is, it can be difficult to attract top graduate students and young researchers, who tend to prefer Beijing, Shanghai, and other more established regions and universities: ultimately, this lack of talent could prove to be a limiting factor for SIP's future growth potential.

Furthermore, several interviewees commented on SIP's prospects for success at becoming China's nanotechnology commercialization hub as requiring the widespread adoption of standard practices for technology transfer. One challenge has to do with the degree to which technologies are in fact ready to be transferred. Having followed the Singaporean model, SIP – like China's other high-tech parks – derives much of its revenue from hosting transnational firms, rather than focusing more directly on domestic R&D that could lead to the eventual commercialization of homegrown technologies. Although an infrastructure in support of nanotech R&D and commercialization has been developed, its payoff in terms of commercial success has yet to be demonstrated. An additional challenge is that while technology transfer is encouraged at present, there is no fixed model for how it should be done, leaving many researchers in the dark about best practices and a path forward. When patents are filed, it is more often done in the US as opposed to China's State Intellectual Property Office (SIPO) – although some interviewees noted at least a desire to begin patenting more in China (SIP interview, April 22, 2012). Nor are these the only problems. At present, there are greater incentives for university and research institute faculty and researchers to publish in high-impact journals, rather than engage in product development. As one interviewee explained, "universities encourage starting up a company, but make tech transfer difficult. Patent transfers are difficult … you can't start a company of your own. The university must approve the company. The university is protective of its IP. The pressures are for professors to do research and teach but not to commercialize" (Shanghai interview, April 18, 2012).

Despite the pressures associated with being small startup companies and labs in an emerging economy, many of the people with whom we spoke commented on their excitement to be in China during this time of unprecedented growth. One engineer with US private sector management experience commented that "every day, the job [in the US] seemed routine, so I came back to China to do more interesting work. Things are very different [in China]. Everyone is talking about nanotechnology – particularly related to energy and biology" (SIP interview, April 25, 2012). Others raised such hopeful sentiments as well. One principal in a small startup firm with whom we met explained that for him, "the most important reason for coming back to China is the chance of getting things right and [having] an extremely successful business" (SIP interview, April 22, 2012). Others noted the growing culture of entrepreneurship as something that attracted them back, while also noting that truly innovative research occurs largely in foreign firms, since "foreign companies pay much more money and therefore, are able to hire the best talent" (SIP interview, April 19, 2012).

Nonetheless, as Chinese firms increasingly shift their portfolios to include R&D in addition to manufacturing, the knowledge spillover will likely boost innovative capacity (Shanghai interview, April 19, 2012). Even with higher salaries, there are significant cost advantages to conducting R&D in China. One interviewee, who has conducted research on China's regional economic growth, estimates that the human capital costs in China are only a sixth of those in the US. These advantages may ultimately favor China's efforts at indigenous innovation (Shanghai interview, April 18, 2012).

Several of the researchers with whom we spoke commented on an increase in international collaboration, both with foreign companies as well as foreign university researchers. Nanotechnology can often demand extremely costly equipment, and the foreign research community has begun to leverage partnerships with their Chinese counterparts to make use of China's investments in infrastructure: "Chinese universities [and labs] have great instrumentation so many foreign companies come to China to use their facilities and for the Chinese students to do research for them" (Shanghai interview, April 19, 2012). There is also a growing realization of the importance of international partnerships for China's nanotechnology commercialization efforts. The head of one startup explained his intention to reach out to international partners for support in marketing, packaging, and branding among other value-added activities after their products were ready to scale up to international markets, explaining that he saw no need to have these capabilities in-house (SIP interview, April 23, 2012).

Throughout our interviews, we heard of many other concerns about impediments to innovation. One commonly voiced worry had to do with overbuilding. One interviewee, for example, questioned whether so much new building space was truly needed: "so many buildings have been built ... [I] haven't seen a lot of people move in so there are a lot of empty spaces" (SIP interview, April 25, 2012). By the same token, we were told that costly machinery and instrumentation is largely going unused by domestic scientists (ibid.). Another concern had to do with cultural barriers that thwart innovation. As one returnee put it, "you can't just construct new buildings. You have to change it from the roots. You need to change the educational system. Creative thinking is needed" (Shanghai interview, April 24, 2012). Employment-related attrition is another concern; because of the high number of new startup firms, and a diminished sense of loyalty to a single firm, young scientists and engineers have a high degree of portability (SIP interview, April 22, 2012). There is also a shortage of experienced managers; as one interviewee commented, "there is more of a need for high-level senior employees who are hard to find and there are few of them in China – there is not much of a need for newly-graduated PhD students" (SIP interview, April 22, 2012). Others also commented on the difficulty of recruiting back top overseas Chinese (SIP interview, April 25, 2012).

Has Suzhou Industrial Park emerged as a nanotech innovator?

As competition for world-class researchers and experienced managers stiffens, and nanotechnology matures as an industry, globally, the hope is that SIP's strategy of concentrating nanotechnology into bioBay and Nanopolis will poise Suzhou for success. Many questions remain, however, as to the extent to which true innovation will occur inside of China's industrial parks and whether these innovations will remain the province of the Chinese nanotechnology enterprise. For example, will foreign firms continue to dominate in nearby Shanghai, or will SIP succeed in luring Chinese-owned firms? Will Suzhou continue to be a compelling destination for returning overseas Chinese with its high quality of life, educational standards, and relative wealth? Or will larger cultural hurdles continue to thwart its success? Will China's education and research cultures that are said to discourage creative thinking hinder the region – and the country more broadly – from becoming a globally competitive nanotechnology leader? And, perhaps most importantly, will China's top-down developmental state approach stimulate genuine innovation, leading to globally competitive high-tech players?

Whether or not Suzhou Industrial Park and its Nanopolis have indeed made the transition to become a true source of cutting-edge scientific and technological development driving indigenous innovation remains to be seen. The long-held promise of nanotechnology has yet to be fully realized in China, as in other countries as well – the US included – which have invested heavily in the technology in the hope of stimulating the next great technological revolution. What may ultimately help SIP and Nanopolis to succeed, however, may be its proximity and ability to leverage nearby Shanghai. This relationship between Suzhou and Shanghai, along with the symbiosis of foreign and local firms (financial and technical), may enable SIP to realize its claim to serve as an important engine of China's quest to become a high-tech innovator.

China's capacity for nanotechnology innovation remains behind that of the US and other advanced industrial economies. As a result, although state policies and funding have had some success in advancing research, they have been less successful in bringing viable products to market. Basic research fails to drive long-term innovation; as a result, new products typically mirror the functionality of existing products, rather than representing breakthroughs – for example, nanoporous materials designed to remove odors from refrigerator surfaces, or hydrophobic materials impervious to spills.

6 Xi Jinping's Chinese Dream
Some Challenges

Xi Jinping's 2012 Chinese dream speech served as a clear indication of China's rising nationalism. His predecessor, Hu Jintao, had championed China as a "harmonious society" by presenting his country as an emerging economy that posed no serious economic or political challenge to the advanced economies of the West, seeking only harmonious social relations both internally and externally. In fact, China was characterized as more or less playing the game according to the West's rules in its Reform and Open Door era so as to not pose a challenge (Steinfeld 2010). Xi's speech, in sharp contrast, can be taken as a shot across the Western bow: it proudly announces China's arrival on the world scene and its intention to rectify the injustices of the past, which had humiliated China and consigned it to a peripheral world status.

Yet Xi has also advanced a less nationalistic, more cosmopolitan vision – one that sees China as a contributing member to a global community, in which all countries pool their knowledge and resources to solve common problems that threaten all of us. A few months before his "Chinese dream" speech, Xi addressed the opening ceremony of the National Assembly of the International Astronomical Union, noting:

> the development of science and technology requires extensive international cooperation. Science and technology have no nationality!... Nowadays the challenges for S&T are more and more globalized, and all humankind is faced with the same problems in energy and resources, ecological environments, climate change, natural disasters,

food security, public health, and so on... Today's world is an open
world, and countries are depending on each other more and more heavily.
(Xi 2012: 1)

On numerous measures – its rapidly rising expenditure on research
and development, its growing high-quality talent pool, its impressive
growth in output of scientific publications and patents – China appears
to be well on its way to becoming a formidable global player, if not yet
a superpower, in science and technology. Behind this increase in activity
lies the highly visible hand of the Chinese state. Since the late 1970s,
the government has issued a plethora of S&T policies designed to
reform the S&T system, to increase investment in S&T and R&D, to
expand the number of scientists and engineers, to establish high-tech
parks, to encourage venture capital investment, to better protect IPRs,
and lately to build a more innovation-oriented nation with strong
international S&T relations. The government has also introduced
industrial policies that support the development of high-tech sectors,
aimed at strengthening industrial competitiveness, encouraging larger
investment in innovation, and promoting high-tech trade. Innovation
financing, preferential tax treatment, and better management of S&T,
R&D, and innovation funds also have become more apparent. Taken
together, S&T, industrial, financial, tax, and fiscal policies have been
integrated to form a steadily more coherent, integrated package of
innovation policies (Liu et al. 2011).

In this chapter we consider some of the implications of China's
changing role on the S&T world scene. We begin with a discussion
of some key challenges China faces in becoming a high-tech economic
superpower, before considering its emerging S&T relations with the
rest of the world. Its S&T system has significant weaknesses; economic
growth has slowed from nearly two decades of double-digit annual
increases; the economy is plagued with high levels of public and private
debt; and there is widespread dissatisfaction resulting from corruption,

pollution, and inequality. We conclude with some speculations about what China's changing role might portend for the United States and, by implication, other advanced industrial economies, given the tension between nationalism and cosmopolitanism reflected in Xi's two speeches.

WEAKNESSES IN CHINA'S S&T SYSTEM

China's S&T system suffers from a number of challenges, some of which we have touched on in previous chapters. Reform of China's S&T system must begin with China's political leadership, which has yet to create an environment suitable for an enterprise-centered national innovation system. In this context, in order to compete in the domestic as well as international markets, enterprises have to not only come up with new ideas and technology but also quickly convert them into new products and new services that generate profit. If they do not have the resources to develop new technology, they likely invest in developing their own human resources or engaging institutions of learning for help. The role of the state in such an innovation system is to invest in education and research systems that serve to support areas where market failure is likely to occur, nurture a culture of innovation, and, most importantly, provide an institutional environment that encourages competition and stimulates and protects innovation and IPRs. In this regard, China may learn from the experience of Taiwan's Industrial Technology Research Institute (ITRI) (Greene 2008). No one can predict which enterprises will become truly innovative, but the cultivation of an institutionally embedded culture of innovation has proven to be more effective than picking winners.

Innovation also requires intermediary organizations that connect the scientific and business communities, by providing necessary information and services (Burt 1992). In general, the ideal of having non-government organizations such as financial, legal, headhunting, accounting, and consulting firms as independent third parties to support technology learning, innovation, and entrepreneurship seems to be the norm in Silicon Valley (Lee et al. 2000). In this regard, the CAS has taken some significant steps by creating innovation and incubation centers in Foshan, Shenzhen, Harbin, and Shanghai to promote the transfer of its scientific research achievements and scale up industrialization. Some CAS institutes also dispatch their staff members with industrial research and technology transfer experience to assist engineers at enterprises in learning and transferring technology. Whether this approach could facilitate and accelerate the transfer of knowledge from laboratory to industry still remains to be seen.

In what follows, we focus on two problems that are particularly vexing: weaknesses in China's STEM educational system, since basic research provides the foundation from which innovative S&T applications eventually emerge, and China's IPR system, which directly affects its ability to effectively turn innovative ideas into viable products.

Chinese higher education

China's higher education challenges begin with its college-entrance exam, the *gaokao*, which largely decides which university a student will attend. In a society where connections outweigh talent and where money outweighs (pretty much) everything, the *gaokao* was supposed to offer a fair way, based on meritocracy, of determining who gets to go to which college. The reality, however, is that not only does it discriminate between students in different cities, but it also discriminates between rural students and urban students. This is mainly because top universities reserve more spots for local students, even if they score lower, or even much lower, than students from other provinces.[1] For example, Peking University and Tsinghua University, together, will accept about 84 students out of every 10,000 Beijing students who took the *gaokao*; 14 out of every 10,000 students from Tianjin; ten out of every 10,000 students from Shanghai; three out of every 10,000 students from Anhui;

and only two out of every 10,000 students from Guangdong and Jiangsu (Fu 2013). This means that for a student from Guangdong to get into Peking or Tsinghua universities, (s)he would have to score significantly higher than someone from Beijing to earn the same spot. In 2014, Jiangsu students had a less than 0.01 percent chance of being accepted into Tsinghua University (Chen 2016).

In a society where inequality is of significant concern (see our discussion below), the unfairness of the *gaokao* and college acceptances strikes a chord with families. In 2016, parents and students protested in at least six cities in Jiangsu province against the unfair treatment of students from different provinces (Chen 2016). Similar demonstrations occurred in Nanjing and Wuhan. Students from top universities such as Peking, Tsinghua, or Fudan are often thought of as the best and brightest from all of China. This may, in fact, not be true given that local students can score much lower *gaokao* scores than their non-local counterparts and still get accepted.

If China hopes to move towards being an innovation-based economy, it needs to find a way of matching its best and brightest students to its best universities. Nor do its higher education challenges end once students find their way into Chinese universities. A key problem was highlighted nearly a decade ago in a controversial (in China) 2010 editorial in *Science*, by Yi Rao and Yigong Shi, then deans of life sciences at Peking and Tsinghua universities respectively, two of China's most prominent returned scientists from the US – Rao from Northwestern, Shi from Princeton:

Although scientific merit may still be the key to the success of smaller research grants, such as those from China's National Natural Science Foundation, it is much less relevant for the megaproject grants from various government funding agencies ... This top-down approach stifles innovation and makes clear to everyone that the connections with bureaucrats and a few powerful scientists are paramount, dictating the entire process of guideline preparation. (Shi and Rao 2010)

The commercialization of public research results has also been rendered difficult because of the view that these results are considered public goods, thus disincentivizing researchers to engage in technology transfer or become entrepreneurs themselves. A 2016 study to better understand the opportunities and challenges of China's STEM research environment at institutions of higher education was conducted using an online survey of STEM research faculty from the top 25 institutions of higher education (Han and Appelbaum 2018).[2] Findings from the study suggested that there remain a number of critical challenges in China's research culture and environment that must first be addressed if China is to become a world-class leader in science and innovation. The study found that the top five challenges to China's research environment, as identified by survey respondents, were that it promotes short-term thinking/instant success, embraces an inequitable research funding process, entails too much bureaucracy and governmental intervention, suffers from an evaluation system that rewards quantity over quality, and involves too much reliance on personal relations (*guanxi*).

The study also found that at the level of basic research, classrooms and laboratories were not always conducive to innovative thinking and scientific breakthroughs. While the majority of survey respondents (55 percent) reported being satisfied or very satisfied with their personal position, nearly half (42 percent) were unsatisfied or very unsatisfied with the overall research culture in China. More than a third complained that the research environment promoted short-term thinking aimed at achieving instant success, rather than a longer-term outlook that could result in important research results.

China's educational culture is based on rote learning and respect for authority, traits that inevitably carry over into its research labs, discouraging innovative thinking. Students follow directions, but seldom think creatively, reportedly choosing their own research topics only 3 percent of the time (data from Han and Appelbaum 2018). On the other hand, it is possible that as a growing number of Chinese expatriate scientists and engineers return to China, attracted both by China's growing global prominence and generous incentives provided by national and local governments, the Chinese research culture may improve as a result. Among survey respondents who held foreign degrees (17 percent of the total), approximately two-fifths were from US universities.[3] Respondents who had studied abroad saw a foreign degree as providing higher quality education and research opportunities, along with a better knowledge of the field.

One critical challenge in China is the evaluation system, with its over-emphasis on quantitative metrics. The consensus among survey respondents was that quantity is strongly valued over quality, and that most researchers are spending more time trying to meet department-wide and university-wide metrics rather than focusing on meaningful or innovative research. In worst-case scenarios, some researchers feel pressed to engage in scientific misconduct to secure the required number of publications. There has been a proliferation of fake papers (Van Noorden 2014); ghostwritten and fraudulent papers (Hvistendahl 2013, 2015); and instances of data falsification, data fabrication, and plagiarism (Gong 2005), as researchers face increased pressures to "publish or perish." Because Chinese universities typically use quantitative metrics such as the number of publications researchers publish each year, and science journal citation indices to determine promotions, salaries, and bonuses (Resnik and Zeng 2010), researchers face mounting pressures to meet these expectations. An associate professor at Fudan University reported that "university management/administration needs a lot of improvement because it is only concerned with the amount of funding received, the number of articles published ... the amount of funding one receives is a very large indicator in one's performance evaluation; people fabricate or plagiarize papers so that they can pass their annual performance evaluations. The same is true for title (i.e., academic rank) evaluations." The dissatisfaction with the current evaluation system among STEM faculty researchers is illustrated by the statement of an associate professor at Sun Yat-Sen University, "only publications are considered important. Evaluation systems are impatient towards researchers – the system requires people to quickly produce research results and products. In the end, it makes researchers very impetuous in their work."

A professor from Tianjin University added that in addition to over-emphasizing quantitative metrics, the system of evaluation "does not care about whether your research actually answers a research problem or not. The result is that it encourages researchers to follow the trend in terms of what they research, and does not support the actual pursuit of knowledge." An associate professor from the University of Science and Technology of China agreed, adding that "overall, the system is impetuous. Where you publish your research is very guided, publishing has become the main objective for why we do research. Meeting publication requirements is directly related to your livelihood, it makes it very hard to do long-term, deep meaningful research."

A professor at the Huazhong University of Science and Technology provided some insight on the specifics of the annual evaluation system at his department. As a professor, he needs to teach a minimum of 32 undergraduate credit hours and a minimum of 32 graduate credit hours each academic year. Additionally, he needs to bring in at least 250,000 yuan (US$38,000) in funding each year and publish at least one SCI publication. When asked what happens if he doesn't meet all of these requirements, he stated that although part of his salary is basic, another part is based on the evaluation criteria,

so if [I] don't meet these requirements, let's say [I] scored an 8 out of a 10 on your evaluation, then [I am] only going to get 80 percent of

[my] "bonus." After taxes, associate professors make slightly less than 100,000 yuan [US$15,000] after taxes. Full professors make roughly about 144,000 yuan/year [US$21,700] after taxes. This is considered pretty low in China [depending on where you are].

He went on to state that new PhD graduates could be making as much as 200,000 yuan a year (US$30,000) or more by going into industry. Companies such as Taobao (one of the platforms of Alibaba), Baidu, and Tencent are paying very well and that a lot of students are choosing to leave academia because of the increased benefits. An associate professor at Wuhan University also commented that more PhD graduates are leaving academia for industry because of salary differences and that this has become a problem for academia in China as in North America.

A professor at Sun Yat-Sen University noted that the evaluation system at his department was a points-based system that considered everything from the number of courses taught to the amount of funding received to the number of graduate students that were mentored and that graduated each year. A point was given for each fulfilled requirement. The professor went on to state the evaluation system causes a lot of strife among professors because of the difference in salary between those who spend more time teaching and those who spend more time on research. He states that a large

> problem is that [a person's] salary isn't steady. If you have a lot of publications one year, you may make a lot of money but next year maybe your salary decreases a lot. You get different points depending on which journal your publication is published in: more points for higher journals, less points for less ranked journals – but how can [the university] state that the impact of my research is less or more than another person's? This has nothing to do with the journal and the system is very unreasonable and impractical.

By reducing scientific research to quantitative metrics, China has decided to place more value on short-term gains over long-term impacts. The study found that the evaluation system is just one of many obstacles that hinders the potential impact of the Chinese science community (Han and Appelbaum 2018).

Weak – but improving – intellectual property rights protection

China lags far behind traditional global S&T powers in terms of patenting. Although China's total patent filing growth has dwarfed other regions, it remains far behind major countries in terms of licensing revenue (Ghafele and Gibert 2012). This is more likely the result of policies that incentivize the filing of patents without adequately emphasizing the importance of patent quality for generating licensing income. These incentive systems precipitated a rush in patent filing, often for inventions of little or no value. The rapid rise in utility-model patents, which do not require inventions to be novel and last only ten years, is a problem in terms of patent monetization. This finding also is supported by the World IP Index 2016 issued by WIPO: Despite being the main driver of world patent and trademark application growth, China lags behind the US, Japan, and other developed countries in terms of patent filing abroad, and applications granted, which attests to its relative low level of internationalization and significant quality gap. In 2014, the world's top three countries in patents filed abroad were the US (approximately 224,400), Japan (200,000), and Germany (105,600), while the number for China was only approximately 36,700, one-sixth that of the US. In terms of patent output standardized to per capita GDP, South Korea produced 2.5 times as many patents as China; Japan, 12.5 times as many (WIPO 2016b).

Despite China's legal efforts to improve IPR protection since joining the World Trade Organization in 2001, there remain numerous hurdles for IPR enforcement. As of 2017, China remains on the

Office of the United States Trade Representative's Priority Watch List because of longstanding concerns related to IP infringement, theft of trade secrets, and high levels of manufacturing counterfeit goods for markets around the world (USTR 2017). For instance, China – along with Hong Kong, India, and Singapore – was associated with 90 percent of all counterfeit pharmaceuticals seized at US borders in 2016 (USTR 2017). In addition, a 2016 report by the OECD and the EU's Intellectual Property Office estimated that the global counterfeit and pirated goods market is worth approximately half a trillion dollars a year, with China as the top producer, and American, Italian, and French brands being the hardest impacted (OECD/EUIPO 2016).

China's IP protection system uses a dual IPR enforcement structure composed of a judicial and an administrative process (Cao 2014). The more commonly used option is the administrative process. Predominantly favored by Chinese companies, this approach is fast, cheap, and simple compared to the judicial process. Foreign firms, however, tend to underutilize this option as it is less familiar than the common approach to IP enforcement in Western countries. For instance, in 2014, China handled 24,479 patent administration cases, of which only 2.1 percent involved foreign rights holders (Covington 2015). In 2016, China saw the number of patent administration cases nearly double from 2014 to nearly 50,000 cases (SIPO 2017). In the administrative process, when a rights holder believes that a patent has been infringed, the case is brought to the local authorities in charge. If the local authority believes that infringement has occurred, an injunction will be issued, often on the same day of the filing (Covington 2015). Local authorities, however, lack the power and authority to award monetary damages. Additionally, they cannot impose sanctions on parties who fail to comply with injunctions. For these reasons, it is often necessary for rights holders to simultaneously file a litigation case under the judicial process for monetary compensation and injunction enforcement.

The judicial process, equivalent to civil litigation in the US and in Europe, is the other IP enforcement option available to rights holders. China's court system consists of four levels: county or district courts, intermediate courts, higher courts, and the Supreme People's Court. Civil litigation is typically brought at first instance to intermediate courts (Duncan et al. 2008). Unfortunately, there are a number of obstacles that make the judicial process less effective than it otherwise could be. One of the biggest impediments to the judicial process is the exceedingly low monetary compensation that is awarded even when courts find that rights have been infringed. In a 2012 study, it was shown that the average amount awarded was 159,000 yuan (US$25,200) with a ceiling of 1 million yuan (US$158,400) (Zhan 2014). As there is no robust system for evidentiary discovery in China (Ong 2009), it is nearly impossible to establish damages that would lead to higher monetary awards.

Chinese courts have also been criticized for favoring local enterprises over foreign entities in a practice known as local protectionism. A common practice within China, local protectionism arises because courts are dependent on the local government for both funding and judiciary personnel appointments (Ong 2009; Lee 2015). Local governments, therefore, often interfere in judicial cases and pressure courts to find in favor of local businesses because they contribute to local economic growth and jobs (Lee 2015). One of the most well-publicized examples of local protectionism is the 2007 case between Chint Group Corp. and a joint venture company owned by Schneider Electric of France. Chint Group Corp. filed a suit against Schneider Electric Low Voltage (Tianjin) Co., Ltd. in its home city of Wenzhou, Zhejiang province, claiming that Schneider's circuit breaker products infringed on Chint's patents. The local judge found in favor of Chint and not only ordered

Schneider to stop producing the products but also awarded Chint an unprecedented amount of 334.8 million yuan (US$44 million) in monetary damages (Duncan et al. 2008; Ong 2009).[4] In patent cases in which the foreign entity is the rights holder, the average amount awarded rarely exceeds a few million yuan (Ong 2009). Despite clear drawbacks associated with the judicial process, however, it remains the primary method in which foreign rights holders seek IP protection in China because it is the only method that offers monetary compensation and enforces injunctions in cases of IP infringement (Yang 2011).

There is reason to believe, however, that as China moves to a more technology-based, innovation-driven economy, it will also heighten IP enforcement within its borders. There is already evidence that suggests China is taking patent infringement cases more seriously, awarding higher monetary damages, and placing more importance on IP enforcement. In 2014, the Supreme People's Court announced the establishment of the much-anticipated Intellectual Property Courts of China. As a pilot campaign, IP Courts have been established in Beijing, Guangzhou, and Shanghai. One noticeable improvement over traditional courts is that judges presiding over IP courts do not have to answer to any higher-level authority (Rotenberg 2015; H. Wang 2017). This change could potentially decrease the level of local protectionism, as judges' decisions should be independent of regional and other external factors or biases. In addition, a recent case in December 2016 indicates that IP Courts may be more likely to award higher monetary damages for IP infringement. Watchdata Co., Ltd. was awarded the highest monetary damages to date by Beijing IP Court of 50 million yuan (US$7.5 million) (H. Wang 2017). As China's IP Courts are still relatively new (the first being the Beijing IP Court, established in November 2014), only time will tell if they will offer more IP protection and enforcement than the current system, and if their practices will be expanded to other cities around China.

FISSURES IN THE CHINESE ECONOMY

China's economy has significant weaknesses that are recognized (and acknowledged) not only by China's critics, scholars, and international institutions such as the World Bank and the European Central Bank, but by China's leaders as well. China's domestic non-financial sector debt is estimated (as of 2017) to be 251 percent of GDP, and is projected to be three times GDP in five years (IMF 2017).[5] China's economic growth has been driven by excessive total investment in such areas as public infrastructure, housing and office building development, relative to private consumption, reaching nearly half of GDP in 2012 (48 percent), although declining to roughly 44 percent since that time (Quandl 2016). China's state-owned enterprises (SOEs), large-scale recipients of public funding and key engines of economic growth, tend to be bureaucratic, risk (and hence innovation) averse, and beholden to party connections. They are also strongly favored when it comes to loans from state-owned banks, relative to more innovative small and medium-sized enterprises.

Economists argue that high rates of investment relative to consumption are untenable in the long run (Lardy and Borst 2013; Pettis 2013). China's leaders acknowledged these problems long before the Xi presidency. As summarized in a posting by the International Economy Division of the Australian Treasury,

[former] Premier Wen Jiabao told a press conference in March 2007 that the economy was "unstable, unbalanced, uncoordinated and unsustainable." Among other causes, the Premier noted that "China's economic growth relies too much on investment and export." The 12th Five Year Programme for National Economic and Social Development, adopted by the National People's Congress in 2011, formally recognizes that there are "imbalanced, incompatible and non-sustainable elements within China's development," which include "an imbalance between investment and consumption". (cited in Hubbard et al. 2012)

While such investment has given China maglev trains, a modern national highway and high-speed railway system, brand new universities, science parks, and an ultra-modern urban facelift in dozens of cities, it has also fueled a bubble economy with miles (and in some cases entire cities) of empty office buildings and vacant apartments (Miller 2012; Gray 2017). Although China's adoption of international banking regulatory norms has reportedly reduced nonperforming loans (NPLs), often issued by local government entities for questionable real estate ventures, from 12 percent of GDP in 2006, to less than 2 percent in 2017 (CEIC 2017), a recent World Bank report concludes that "without reforms in current lending practices and governance structures of state-owned commercial banks, the banks are likely to again accumulate large NPLs if the economic growth rate slows and the performance of SOEs deteriorates" (World Bank 2013: 126).[6]

There is considerable debate over whether or not China will succeed in rebalancing its economy – transitioning from public investment to private consumption, reining in questionable local investments in infrastructure and housing, and redirecting economic growth from low value-added production to high value-added technological innovation.[7] Will China join other emerging economies and succumb to the so-called "middle-income trap"[8] (Gill and Kharas 2007), as its economy slows? In the view of one study that statistically modeled determinants of the middle-income trap,

> slowdowns are more likely in economies with high old age dependency ratios, high investment rates that may translate into low future returns on capital, and undervalued real exchange rates that provide a disincentive to move up the technology ladder. These patterns will presumably remind readers of current conditions and recent policies in China, the case motivating much of the slowdown literature. (Eichengreen et al. 2013: 12)

On the plus side, the study also notes that China has slightly higher average years of secondary schooling, as well as a higher share of exports that are high-tech products, than do the "trapped" cases in its analysis.

The World Bank *World Development Report 2017* examined the middle-income trap, and identified a number of structural sources of the problem. Surplus agricultural labor that initially provided a ready source of low-wage manufacturing labor becomes exhausted; labor shortages then drive manufacturing wages up, rendering export-oriented industrialization less competitive and less viable as a growth strategy. High levels of income inequality, which result in social divisions that prevent viable growth coalitions from emerging, is another problem. A third challenge occurs when productivity lags – for example, when more productive, technologically advanced firms fail to displace less productive ones. The World Bank report also identified what it characterized as "political economy traps," when entrenched interests that benefited from the growth spurt into middle-income status thwart changes that they perceive as threatening:

> One such political economy trap is a persistent deals-based relationship between government and business. Deals-based, sometimes corrupt, interactions between firms and the state may not prevent growth at low income levels; indeed, such ties may actually be the "glue" necessary to ensure commitment and coordination among state and business actors... But they become more problematic for upper-middle-income countries. (World Bank 2017: 160)

All of these are challenges that China currently faces. The ability of China's leadership to successfully manage the transition into leading-edge technologies will play a central role in China's ability to become a global economic player, whereby its citizens may move into higher income status.

PUBLIC DISSATISFACTION

The Chinese have mixed and often conflicting views on the rapid changes that are transforming their country and traditional way of life. Issues ranging from toxic air quality to widespread corruption have led to substantial capital outflows from China, as the holders of newly-accumulated wealth seek to hedge their bets on China's economic future, putting their money in foreign investments – or simply squirreling it away in foreign securities (Liu 2017). Wealthy Chinese are reportedly also considering relocating themselves (as well as their money), with one survey finding that half of China's millionaires are considering emigration – with the US and Canada as the most popular destinations (Tencer 2017). These concerns are widely shared, including by the scientists, engineers, and entrepreneurs who are key to China's global S&T ambitions.

On the plus side, a survey conducted by the Pew Research Center in 2015, which sampled 3,649 adults,[9] found virtually universal satisfaction with the overall economic situation: 96 percent reported that their standard of living was better than that of their parents at the same age; 77 percent reported being better off financially than they were five years earlier; and 72 percent claimed their personal economic situation was good. Although fully two-thirds claimed they liked the pace of modern life, nearly four out of five also believed that their way of life needed to be protected against foreign influence. Fully two-thirds felt their traditional way of life was getting lost, and slightly more than half felt that consumerism and commercialism was a threat to their culture (Wike and Parker 2015).

There is, however, evidence that discontent is more pervasive than suggested by survey results. The Chinese government began publishing statistics on "mass incidents" (those involving more than ten people) in the mid-1990s. The reported numbers grew steadily, from an estimated 10,000 in 1994 to more than 100,000 in 2008, which would suggest several million or more people were involved in that year. 2008, however, was the last year in which official statistics were made available, leaving estimates to activists who rely on the Internet, including websites and social media, to track protests, riots, strikes, and public disturbances that are sufficiently large to attract attention (Wu 2016). One prominent blogger, Lu Yuyu, estimated that in 2015 there were nearly 29,000 incidents involving ten or more people, representing a one-third increase from the previous year. These included as many as 40 protests a month involving more than a thousand people.[10] Such numbers suggest not only widespread and pervasive discontent, but also that protesters are often willing to take to the streets in large numbers despite official repression of public demonstrations. Grievances include wage and other workplace violations, land expropriation and forced demolitions, police violence, equal access to education, and even the taxi system (Wu 2016). The above-mentioned Pew survey (Wike and Parker 2015) identified several sources of discontent that are viewed as either a "very big" or "moderately big" problem by at least three-quarters of respondents: graft and corruption, environmental degradation, and income inequality.

Graft and corruption, which permeate the economic and political system at all levels, is the leading source of resentment among Chinese citizens. The Pew survey found that corrupt officials were the top concern of respondents, with 84 percent agreeing that this was either a "very big problem" (44 percent) or a "moderately big problem" (40 percent). President Xi has long been aware of these problems, and in 2012, in his inaugural address as General Secretary of the Communist Party, he identified fighting corruption as a central challenge for the party. He subsequently launched a far-reaching anti-corruption campaign, which has enjoyed popular support: the Pew survey reported that nearly two-thirds of respondents believed the campaign would reduce corruption over the next five years. By some measures the campaign has been successful: it has netted more than a million low- and mid-level public officials, business tycoons, as well as some leading members of the Communist Party (Associated Press 2016). Much of the punishment for graft and corruption has reportedly been mild, however, despite some high-profile figures who received severe prison sentences. During the period 2013–15, for example, of 7.4 million party members in the civil service, roughly one out of every ten was disciplined; most received only warnings or demerits. Fewer than 40,000 were punished for graft and expelled from the party (Lockett 2016). As we have noted above, corruption in scientific research, including but not limited to misuse and abuse of research grants, jeopardizes China's efforts at innovation.

Whatever its success, the anti-corruption campaign has not necessarily shored up the legitimacy of the central government. According to a survey of 83,300 people conducted by the Institute of Governance and Public Affairs of Guangzhou's Sun Yat-sen University, the larger the number of local officials targeted for corruption, the more likely were residents of the locality to believe that corruption was even worse in Beijing (Ni and Li 2016). Moreover, while the anti-corruption campaign is intended to capture both "tigers and flies" (*laohu cangying*) – or top officials and low-level functionaries, it has also not escaped notice that the tigers include some of President Xi's political rivals (Griffiths 2016).

Environmental degradation is also a well-known problem, as well as another major source of discontent: the Pew survey found air pollution to be the second most cited concern of respondents, with 76 percent regarding it as either a "very big" or "moderately big" problem; water pollution was third (75 percent). State-owned firms sometimes dump toxic chemicals into rivers, are a major source of air pollution, and are often resistant to the adoption and enforcement of higher standards. Lancet's *Global Burden of Disease Study 2010* estimated that air pollution accounted for 1.2 million premature Chinese deaths in that year, two-fifths of the world's total. Government policies such as the 2013 Atmospheric Pollution Prevention Plan[11] have since reduced air pollution to some extent, although as of 2016, small-particle pollution remained more than three times the World Health Organization's recommended level in the Beijing–Tianjin–Hebei region;[12] four out of five cities that monitored air quality failed to meet national standards in 2015. In some cities, small-particle pollution reached critical levels: Shenyang, for example, has achieved levels 60 times higher than the WHO's safety standard (Dong 2015; Ouyang 2017). Unbreathable air may drive talent away from China, while discouraging promising expats from returning (Galbraith 2014).

The Chinese Academy of Environmental Planning put the cost of ecosystem damage at nearly a quarter of a trillion US dollars in 2010, equivalent to 3.5 percent of China's GDP (Albert and Xu 2016). China's 2015 Environmental Protection Law has since enacted harsh punishment for polluters, and Chinese investment in green energy is estimated to be double that of the US (Jaeger et al. 2017).[13] But the country's continuing growth makes it unlikely that its environmental problems will be easily solved: car ownership, for example, is predicted to more than double, reaching 390 million vehicles in 2030 (Lancet 2017; Ouyang 2017).

Income inequality is another source of widespread discontent. China's growth has resulted in vast inequality, reflected in one of the largest gaps between rich and poor in the world. While the United States continues to claim the largest number of billionaires with 565, China ranks second with 319, and is closing the gap: in 2017 the PRC posted the largest number of new entrants to the global billionaires list (Forbes 2017). The ostentatious display of extreme wealth showcases a lifestyle that many Chinese emulate but few can hope to achieve, since opportunities for upward mobility continue to be limited, as family background significantly determines an individual's life course.

There are some indications that beginning in 2010–12, inequality began to decline. The most comprehensive recent study of inequality drew on multiple sources[14] to examine trends over the 20-year period

1995–2014 (Kanbur et al. 2017). The study found that based on one widely used measure of income inequality, China's Gini coefficient increased from 0.35 in 1995 to 0.53 in 2010, before declining slightly to 0.50 four years later.[15] By another common indicator, the income share of the top 10 percent was 4.8 times that of the bottom 10 percent in 1995, growing to 19.9 times greater in 2012, after dropping slightly (to 19.1) in 2014. Disparities in wage income, the study also found, accounted for a large (and growing) contribution to trends in inequality, as waged employment grew in importance in the Chinese economy.[16] This was taken as a potentially positive sign, since wage disparity has actually declined slightly over the 20-year period, one possible reason for the noted lessening of overall inequality in recent years (Kanbur et al. 2017: tables 3.1B, 3.2B, 4.1, and 4.2).

The study also conducted a longer-term examination of income trends, beginning in 1978, when Deng Xiaoping began to liberalize the Chinese economy. The opening up of agriculture and industry, which began slowly and experimentally at first, was associated with declining regional inequality for the first few years. Beginning in the early 1980s, however, regional inequality increased sharply as Deng's reforms ("Socialism with Chinese characteristics") took hold, before beginning a decline around 2005. Urban–rural differences played a somewhat larger role in shaping regional inequality at the beginning of this period, with coastal–inland differences becoming more important at the end. On the positive side, both urban–rural and coastal–inland inequality have shown modest declines over the past 10 years. This hopeful trend is in part attributed to tightening labor markets throughout China, as well as state policies such as rural medical insurance and social security, promotion of development in western China, and labor laws that to some degree assure urban workers higher wages and greater job stability (Wang et al. 2009).

Recent marginal improvements notwithstanding, middle-class Chinese continue to struggle with rising living costs and the absence of a viable support net to replace what was once the iron rice bowl tied to one's *hukou* (household registration system), and hundreds of millions of rural farmers remain marginalized from China's economic boom. Perhaps this is why income inequality ranks fourth on the Pew survey (only slightly behind corruption, and tied with air and water pollution), with 75 percent of all respondents regarding the "gap between rich and poor" as a "very big" or "moderately big" problem.

THE GREAT UNCOUPLING: SOME SPECULATIONS ABOUT FUTURE CHINA–US RELATIONS

China's future economic growth is dependent on its ability to compete at the top of the value chain, where success depends on harnessing advanced S&T to innovative products and services. This means that Chinese brands will have to be competitive, both globally and to China's rapidly expanding middle class. China's development strategy calls for shifting away from dependence on state-led investment to private consumption. If this strategy is to succeed, Chinese products must increasingly be designed, marketed, manufactured, and sold by Chinese firms to Chinese consumers, rather than depending on foreign multinationals that employ Chinese workers to make products that are sold elsewhere.

If China succeeds in its ambitions, it could have significant foreign policy implications. At present, China and its trading partners – most significantly the United States – are highly interdependent. The US trade deficit with China accounts for a significant portion of China's foreign account surplus, a *quid pro quo* that has proven extremely beneficial for both China and the US, despite constant grumbling and frequent protests on both sides. China's trade surplus with the US helped to finance the massive investments in science, technology, and infrastructure that contributed to its rapid economic growth. For the

US, China's trade surplus has kept inflation low by providing consumers with an endless stream of low-cost goods, bolstered the revenues and profits of US corporations, and pumped capital into the US economy through China's use of its foreign exchange to purchase treasury securities.

As of June 2017, China (excluding Hong Kong) held US$1.1 trillion in US treasury securities, accounting for slightly more than a third of China's foreign exchange reserves, and nearly a fifth of all US treasuries held abroad (Shenzhen Daily 2017; US FRB 2017a).[17] China's large holdings have given it a vested interest in the continued stability of the dollar; the US has benefited from China's willingness to serve as a major financier of its federal deficit. At the same time, this mutually beneficial relationship has had its downside as well: for China, it has contributed a vast supply of funding for speculative investments in housing, office space, and infrastructure that may prove to be a bubble; on the US side, there has been a loss of manufacturing jobs.

This at-times dysfunctional yet mutually beneficial relationship may well come to an end as China becomes less dependent on foreign technology, foreign multinationals as a source of Chinese jobs, and foreign consumers for Chinese-made products. If China is successful in its quest for indigenous innovation and a consuming middle class, its leaders will no longer see their country's fate as tightly coupled with that of the US as they have in the past. US markets need no longer be a principal engine of economic growth; the dollar no longer a principal source of public funds. While China cannot uncouple its economy from that of the US precipitously without jeopardizing the value of its dollar reserves, in the long run just such an uncoupling may well occur, as China moves up the value chain – increasingly designing and marketing its own high-technology products, selling to its own growing internal market, and offshoring its low-cost, low-wage manufacturing to Vietnam and impoverished countries in Africa. The BRI project in particular – if successful – will link China to its own periphery, a source of low-cost labor and raw materials, as well as a new outlet for its goods and services.

As indicated at several points in this book, we may expect China to increasingly act as a great power, using its economic and political influence to shape world events in its own interest. China will become an S&T exporter rather than importer, extending its S&T relations with emerging economies in Asia, Latin America and Africa – perhaps at the expense of S&T relations with the US and other advanced industrial nations. Its dollar reserves would diminish, and China would press – with eventual success – for the yuan to become a truly global currency reserve.[18] Under the most optimistic version of this scenario, China follows the path of other East Asian economies: it liberalizes its economy, with greater political liberalization following. China emerges on a par with the US, Europe, and Japan as an S&T leader and global economic powerhouse (see, e.g., Glänzel et al. 2008). In short, it pursues a path of globalization consistent with the Bretton Woods framework, seeking a dominant role while conforming to World Trade Organization requirements.

Under a less optimistic scenario, China establishes a new path to "globalization with Chinese characteristics" (Zukus 2017), with such features as central planning and state-led industrial policies; the dominant role of state-owned enterprises in key sectors (such as energy), projected abroad as a form of transnationalized state-owned corporate power; and an authoritarian state, in which the Communist Party's power is asserted politically, economically, and culturally (Henderson et al. 2013).[19] Uncoupled from the US, China flexes its economic and political muscle, with greater freedom to seek to shape world events in its own interests. It develops synergistic (seen by some as hegemonic) relations with emerging economies, providing infrastructure and investment while extracting wealth from cheap labor as well as natural resources. To enhance its growing global power, China develops its military: its navy and air force to protect trade routes and secure itself against depredation from regional rivals; the People's Liberation Army to secure Chinese interests abroad while assuring quiescence at home. Geopolitical rivalries reassert themselves, and a wave of protectionism once again threatens global stability.

Will the Middle Kingdom, after conceding technological, economic, and political leadership to Western powers for centuries, reassert what it has come to regard as its historic place among nations, pursuing a path of global dominance? Or will China work with other countries to help sustain global economic growth, addressing such common challenges as global climate change? At least part of the answer will rest with the US and other leading powers. If the US seeks to punish China in response to its own economic challenges, it seems likely that China will respond both defensively and assertively. As we noted in chapter 4, the US Chamber of Commerce was quick to dismiss indigenous innovation as a "web of industrial policies" in the service of large-scale technology theft (McGregor 2010). China's critics have described its state-led approach to economic development as "innovation mercantilist … a threat not only to the US economy, particularly its advanced industries, but indeed to the entire global economic and trade system," since

> despite the claims of some apologists for Chinese behavior, it's clear what the end game is: Chinese-owned companies across a range of advanced industries gaining significant global market share at the expense of American, European, Japanese, and Korean competitors. (Atkinson et al. 2017: 1)

These arguments have found favor with the Trump administration, which has accused China of currency manipulation, unfair import duties, protectionism, and intellectual property theft (not to mention making up the notion of climate change) – and has vowed to bring China to heel. China, in turn, is concerned that the US is mainly seeking to maintain its superpower status, in the face of China's economic and geopolitical ascendance.

China's advances in science and technology can be seen as a counterbalance to America's once unquestioned technological dominance – or as an opportunity for a global science and engineering effort to solve common problems through expanded cross-border cooperation. In contrast to the past – when bureaucratic rigidities and lack of global awareness limited Chinese options – today China's growth trajectory seems adaptable to the possibilities. How it plays out geopolitically in the future will depend in large part on how Chinese leaders perceive the behavior and responses of other countries within the context of an increasingly fluid political situation, including the possibility of trade wars as well as accelerated innovation processes, shorter product life cycles, and potentially game-changing technological breakthroughs in fields such as artificial intelligence that could significantly alter the global balance of power in science and technology in the years ahead.

Notes

Introduction: From the World's Factory to the World's Innovator?

1 The IMF (2014) estimated a US GDP of $17.4 trillion in 2014; China (adjusted for purchasing power parity) at $17.6 trillion.

2 China's middle class (defined by the OECD as households with daily spending between $10 and $100 per person, adjusted for purchasing power parity) is estimated by the OECD at 157 million people; based on one 2010 OECD projection, it was expected to increase to as many as 840 million people by 2020 (Kharas 2017: figure 6). More recently, one Brookings Institution estimate sees China as adding more than 300 million people to the global middle class between 2015 and 2022 (Kharas 2017: figure 7), which would bring the total to as many as 500 million by that year. While such projections are necessarily based to a large extent on *ceteris paribus* assumptions, barring unforeseen economic changes, China's middle class will continue to grow rapidly in the foreseeable future.

3 Wages in the Pearl River delta, for example, increased nearly 30 percent between 2010 and 2013. In Shenzhen, the monthly minimum wage increased 36 percent between December 2010 and June 2012, from $173 to $236 (Orlik 2013). Many regional authorities mandate annual increases in the minimum wage; by 2017, monthly wages in eastern China ranged from approximately 1,300 yuan (US$200) in Guangdong province to 2,300 yuan (US$350) in Shanghai (Koty and Qian 2017).

4 China's expenditure on research and development started to increase in 1995 when rejuvenating science, technology, and education was determined to be a new development strategy.

5 *"State Council Issue of the 12th Five-Year Plan: Building Programs for National Indigenous Innovation Capacity Notice,"* dated January 15, 2013 and uploaded

to the Chinese government website on May 29, 2013. The Chinese version is available at http://www.gov.cn/zwgk/2013-05/29/content_2414100.htm.

6 New materials include "new functional materials, advanced structural materials, high-performance composite materials, membrane materials, organic silicon materials, nano-materials, and common basic materials."

7 The notice calls for advances in new energy vehicles, including "plug-in hybrid vehicles, pure electric vehicles, fuel cell vehicles, car battery, drive motor, and powertrain."

8 This description is usually reserved for Beijing's Zhongguancun; that notwithstanding, see the Oxford University interview with SIP Chairman Zhiping Yang, titled "China's Silicon Valley – Suzhou Industrial Park," extravagantly described as "one of the most rapidly developing areas in China and around the world." Voices from Oxford Video Interview (November 23, 2012) (http://vimeo.com/54132940).

Chapter 1: China's Science and Technology Policy: A New Developmental State?

1 Among the 386 firms PwC surveyed, roughly half were wholly owned foreign multinationals, a quarter were Chinese companies, and a tenth were joint ventures (the remaining firms did not disclose the relevant information). The top 10 Chinese companies (in rank order) were Huawei, Tencent, Alibaba, Xiaomi, Lenovo, Haier Electronics, Baidu, Byd Auto Co., Meizu Technology, and China Merchants Bank (Veldhoen et al. 2014: Exhibit C, p. 27).

2 BCG's ranking (2017a: 15; 2017b, 2017c) is based on a survey of more than 1,500 senior executives "who represent a wide variety of industries in every region worldwide," and includes respondents' overall picks, along with the selected financial indicators: total shareholder return (stock price appreciation and dividends), revenues, and margin growth. The top companies (in rank order) are Apple, Google, Tesla, Microsoft, Amazon, Netflix, Samsung, Toyota, Facebook, and IBM.

3 The 12 major indices (termed "pillars") include "institutions, infrastructure, macroeconomic environment, health and primary education, higher education and training, goods market efficiency, labor market efficiency, financial market development, technological readiness, market size, business sophistication, and innovation." These pillars are in turn organized into three subindexes: "basic requirements, efficiency enhancers, and innovation and sophistication factors" (WEF 2016, 4).

4 On several measures arguably related to innovation, China also comes up short: education and skills (74th), current workforce skills (85th), and future workforce skills (58th) (WEF 2016: Table 2).

5 The index and its subcomponents are derived from available statistical measures. Overall innovation is calculated on the basis of a formula that considers both "innovation inputs" (measures of the institutional framework, human capital and research, infrastructure, market sophistication, and business sophistication), and "innovation outputs" (measures of knowledge and technology outputs and creative outputs). Details can be found in Dutta et al. (2016: 14).

6 This ranking reflects the fact that China received top scores in indicators such as patent applications by origin, utility model applications by origin, high-tech exports, creative goods exports, global R&D companies, domestic market scale, research talent in business enterprise, and industrial designs by origin (Dutta et al. 2016: 10).

7 "Shanghai is bound to remain among the world's leading innovation hubs given its growing base of digital media and entertainment companies and a more pleasurable lifestyle and favorable climate that can draw top talent. Beijing continues to be regarded as a top leading tech hub ranking third" (New York City ranked second; Tokyo tied with Beijing for third) (KPMG 2017: 5). It should be noted that such optimistic conclusions reflect at least in part the weight of Chinese high-tech executives in KMPG's surveys.

8 Responses were generally regionally skewed. Among all North American (US and Canadian) respondents, 46 percent favored the US, while only 18 percent favored China; among all Asian respondents (China, India, Korea, Japan, Taiwan, Singapore, and Australia), the corresponding percentages were partially reversed (35 percent favored China, while only 13 percent favored the US); among European, Middle East, and African respondents (UK, Netherlands, Germany, Israel, Russia, and South Africa), 27 percent favored the US, 22 percent the UK, and 18 percent China.

9 The Plan also calls for extensive development of oil and gas fields and coal bed gases; large-scale advanced nuclear power plants; control and treatment of water pollution; creation of genetically modified organisms that are resistant to insects, diseases, drought, and cold; development of major new drugs for malignant tumor, cardiovascular, and cerebrovascular diseases; prevention and treatment of infectious diseases such as HIV/AIDS, hepatitis B, and pulmonary tuberculosis; various aircraft and space exploration projects; and the development of new energy (hybrid, electric, and fuel cell) vehicles.

10 China's patents have also increased significantly (see chapter 3). China's patent office – the State Intellectual Property Office (SIPO) – is the largest in terms of the number of patent applications received annually. Nonetheless,

as noted in an OECD (2017a) report, China continues to face challenges in technology transfer from university to firm.

11 Social science and health science publications were excluded from all analyses. Only publications that were considered to be in the physical or life sciences categories as classified by Scopus were included in our analyses. Specifically, the subject areas that were excluded from our analyses are Medicine; Social Sciences; Arts and Humanities; Business, Management and Accounting; Psychology; Nursing; Economics, Econometrics and Finance; Health Professions; Multidisciplinary; Veterinary; Dentistry; and undefined.

12 The Nature Index tracks affiliations of high-quality nature science articles and charts publication productivity for institutions and countries.

Chapter 2: Science and Technology in China: A Historical Overview

1 Research institutes affiliated with the Academia Sinica were Geology, Astronomy, Meteorology, Social Sciences, Physics, Chemistry, Engineering, History, and Language (founded in 1928), Psychology (1929), Zoology, Botany, and Pharmaceutics (1944), and Mathematics (1947). Research institutes affiliated with the Peiping Academy included Physics, Chemistry, Zoology, Botany, and History (1929), Medicine (1932), Physiology (1933), and Crystallography and Atomics (1948).

2 "Black" class backgrounds were designations used to identify groups believed to be enemies of the revolution, and therefore in need of revolutionary re-education (often following public humiliation and sometimes physical beatings). The five black class categories were landlords, rich farmers, counter-revolutionaries, rightists, and "bad elements."

3 This is research supported through the National Natural Science Foundation of China.

4 This includes presumably the mega-engineering programs under the Medium- and Long-Term Plan for the Development of Science and Technology (2006–20) (MLP) and programs administered by different government agencies.

5 This combines the 863 Program, the 973 Program, State Key Technologies R&D Program (the *zhicheng* Program), the mega-science programs under the MLP, other programs administered by the MOST, national S&T programs of public goods run at 13 government agencies, as well as special industrial development R&D programs administered by the National Development and Reform Commission (NDRC) and the Ministry of Industry and Information Technology (MOIIT).

6 These integrate programs of technological innovation at the NDRC, MOIIT, MOST, and the Ministries of Finance and Commerce.

7 These include programs for national laboratories under various names administered by the MOST, the Ministry of Education (MOE), and the NDRC.

8 As mentioned earlier, in 2015, Tu Youyou became the first mainland Chinese to win the Nobel Prize for Physiology or Medicine for her research on artemisinin as a cure for malaria in the 1960s and 1970s.

Chapter 3: China's Science and Technology Enterprise: Can Government-Led Efforts Successfully Spur Innovation?

1 It is not clear how this leading group relates to the State Leading Group for Science, Technology, and Education except that members of the two groups overlap a great deal.

2 These words are taken from Jiang Zemin's report to the CCP's 15th Congress which reads, "We should formulate a long-term plan for the development of science from the needs of long-range development of the country, taking a panoramic view of the situation, emphasizing key points, *doing what we need and attempting nothing where we do not*, strengthening fundamental research, and accelerating the transformation of achievements from high-tech research into industrialization" (emphasis added). In the May 1995 decision, the wording was slightly different: "catching up what we need and attempting nothing where we do not."

3 Strategic emerging industries include energy-efficient and environmental technologies, next-generation information technology, biotechnology, high-end equipment manufacturing, new energy, and new materials.

4 Nearly two decades ago, Unger and Chan (1995) characterized China's reform and transformation as ultimately state-led. Huang (2008) makes a similar argument.

5 For a discussion on picking winners and "manufacturing stars" (the creation of national champions by the government) as "a path-dependent legacy of the postwar developmental state," see Wong (2011: 133–9).

6 One comparison with Taiwan argues that the marriage of political and scientific leadership may hurt rather than promote high-tech development in China (see Greene 2008).

7 Interviews with China's S&T policy analysts and government officials.

8 For a discussion on this in Korea, Taiwan, and Singapore, see Wong (2011).

9 The Taiwanese experience suggests that the state should focus on creating policies for the private sector that foster innovation (see Greene 2008). For a discussion

on the particular role of the state in the development of biotechnology in Korea, Taiwan, and Singapore, see Wong (2011).

10 One of China's top computer scientists, Chen Jin, became famous in 2003 when he claimed to have developed a technologically advanced microchip that rivaled those from more technologically advanced countries. Three years later, the Chinese government revealed that Jin had faked his reported research at Jiatong University and stolen the chip designs from a foreign company (Barboza 2006). The faked chip was developed with funds from several Chinese government agencies (the Ministries of Science and Technology [MOST] and Education, the National Development and Research Commission, and the Shanghai Municipal government).

11 This conclusion draws from Cao et al. (2009).

12 SCI is a bibliometric database compiled by Clarivate Analytics, a company formed after Onex Corp. and Baring Private Equity Asia acquired the intellectual property and science assets, including *SCI*, from Thomson Reuters.

13 The report also found that the top three corporations in terms of patents granted were State Grid (3,622), Huawei (3,293), and Sinopec (2,567). In terms of patents per 10,000 persons, the top three provinces were Beijing (94.5), Shanghai (41.5), and Jiangsu (22.5). In terms of regional innovation, Guangdong province is ranked first, however, in terms of R&D spending, growth in the number of high-tech enterprises, and its plans to expedite the construction of the Pearl River Delta National Indigenous Innovation Demonstration Zone (*China News* 2018).

Chapter 4: China's International S&T Relations: From Self-Reliance to Active Global Engagement

1 Outside of China, it is often referred to as the One Belt One Road Initiative.

2 For example, the "Opinions of the CCP Central Committee and State Council on Deepening S&T Reform and Speeding Up the Building of a National Innovation System," the 13th Five-Year Science and Technology Plan (discussed in chapter 1), the Innovation-Driven Development Strategy, and the Belt and Road Initiative on building international S&T cooperation networks.

3 Others include the Foreign Affairs Leading Group of the CCP, and the Inter-Ministerial Coordination Mechanism, which includes the Ministry of Agriculture (MOA), Ministry of Education (MOE), the international cooperation departments of local governments, the China Association for International Science and Technology Cooperation, and enterprises. MOST also commands some 20 affiliated agencies, including the Institute of Scientific and Technical

Information of China, the High-Tech Research and Development Center, the Intellectual Property Rights Center, the Supervision Service Center for Science and Technology Funds, and the National Science and Technology Venture Capital Development Center.

4 See in particular the *13th Five-Year Plan Special Program on International S&T Cooperation*.

5 "Plan S&T opening and reform with global vision, focus on research areas that receive extensive attention by S&T community and will have significant impact on development and progress of humanity, pool top tier S&T resources from domestic and abroad, strive to make a series of landmark R&D achievements, and improve China's innovation capability in strategic frontier as well as international influence."

6 Information in this section on Japan and China S&T cooperation is mainly derived from assorted issues of the JETRO China Newsletter, 1980–2000 (Japan External Trade Organization, Tokyo).

7 Information in this section is largely drawn from the website of the Delegation of the European Union to China (https://eeas.europa.eu/delegations/China_en).

8 For detailed facts and figures regarding priority areas of FP7 and H2020 as well as Chinese participation, refer to http://ec.europa.eu/research/iscp/pdf/policy/1_eu_CHINA_R_I.pdf#view=fit&pagemode=none

9 The Patent Cooperation Treaty (PCT) provides an international legal framework intended to insure standard patenting procedures that provide IP protection.

10 For a somewhat different conclusion, see Baark (2014), who argues that China does not yet possess the excellence that positions its scientific research institutions as world-leading, even if research in key organizations may be able to support leading and original research achievements.

Chapter 5: How Effective Is China's State-Led Approach to High-Tech Development?

1 China's approximately 150,000 SOEs (a third of which are owned by the central government, the remainder by local governments) account for 30–40 percent of GDP (export.gov 2017; OECD 2017b). Seventy-seven of the 109 Chinese companies listed on the Fortune Global 500 for 2017 were wholly or in large part SOEs, with three Chinese energy-related SOEs among the global top four (State Grid, Sinopec, and China National Petroleum) (Fortune, 2018).

2 Management support includes "hands-on management assistance, access to financing and orchestrated exposure to critical business or technical support services … shared office services, access to equipment, flexible leases and expandable

space – all under one roof.... Industrial estates ... generally have a non-selective intake, provide little or no management support and have no special criteria with regard to business activities and technology content. At the opposite extreme ... technology centres have highly selective admission criteria, provide "hands-on" management support, and have a highly specialized technology focus (European Commission 2002: 5–6).

3 The top 10 unicorns for 2016 are: (1) Uber (US), (2) Xiaomi (China), (3) Airbnb (US), (4) Palantir (US), (5) Didi Kuaidi (China), (6) Snapchat (US), (7) China Internet Plus (China), (8) Flipkart (India), (9) SpaceX (US), and (10) Pinterest (US) (Fortune 2016).

4 Shenzhen High-Tech Industrial Park was established in Nanshan District in 1996.

5 Tencent is also an investor in Didi and Didi has invested in Southeast Asian countries.

6 Research conducted by the University of California-Santa Barbara's Center for Nanotechnology in Society under NSF award SES 0938099 (http://www.cns.ucsb.edu/; see especially http://www.cns.ucsb.edu/research/irg2).

7 Nanotechnology involves working with materials at a scale of less than 100 nanometres (Roco 2007: 3). This provides "the ability to work at the molecular level, atom by atom, to create large structures with fundamentally new molecular organization," allowing materials and systems to "exhibit novel and significantly improved physical, chemical, and biological properties, phenomena, and processes due to their nanoscale size" (NSTC 2000: 19–20). This scale is comparable to the smallest virus, 80 nm, and the diameter of human DNA, about 2.5 nm.

8 There have been a number of Nobel Prize awards related to nanotechnology. In Physics these include development of the space–time view of quantum electrodynamics by Richard Feynman (1965), the discovery of the quantized Hall effect by Klaus von Klitzing (1985), the design of the scanning tunneling microscope by Gerd Binning and Heinrich Rohrer (1986), the discovery of a new form of quantum fluid with fractionally charged excitations by Robert Laughlin, Horst Stormer, and Daniel Tsui (1998), and the discovery of giant magnetoresistance by Albert Fert and Peter Grünberg (2007). In Chemistry these include the discovery of C60 (better known as fullerenes) by Robert Curl, Harold Kroto, and Richard Smalley (1996) and the discovery and development of conductive polymers by Alan Heeger, Alan MacDiarmid, and Hideki Shirakawa (2000).

9 There are military uses as well, which we did not explore in our research. These include nano-electro-mechanical sensors capable of detecting and identifying a single molecule of a chemical warfare agent, and nanocomposite materials that create propellants and explosives with more than twice the energy output of typical high explosives.

10 Of these firms, 1,000 were from Germany and 600 from Japan; China had yet to show a significant presence (US NAS 2016: 12).

11 The exception was Foxconn, a Taiwanese electronic manufacturing giant that does most of its production in China and established the Tsinghua–Foxconn Center of Nanotechnology at Tsinghua University in 2003 (Cao et al. 2013a).

12 The US National Academy of Sciences notes that it typically takes from 10 to 20 years for an initial discovery to result in viable commercial products (US NAS 2016: 14).

13 While Rodríguez-Pose and Hardy (2014) focus primarily on Beijing's Zhong-guancun Science Park – the one example that clearly satisfies their four criteria – they also note that Suzhou Industrial Park is now regarded as an "emerging technology park," thanks in part to technology transfer from its Singaporean and US multinational tenants to Chinese firms.

14 A prefectural-level city is an administrative unit typically composed of several counties, the central city bearing its name, smaller cities and towns, and rural areas. While the urbanized city of Suzhou has a population of roughly 5.5 million, the prefecture is home to nearly twice as many (10.5 million).

15 Jiangsu province has a population of roughly 79 million. Its capital is Nanjing, whose status as a sub-provincial city provides a higher administrative ranking than Suzhou's prefectural-level city status.

16 Of SIP's 111 square miles, the China–Singapore Cooperative Zone occupies roughly 31.

17 The High-Tech District is administered solely by the Suzhou government, is supported by MOST as a National-Level High-Tech Park, and is home to the headquarters of most of Suzhou's major banks. As a result, it has become a "fierce competitor" of SIP.

18 The development zones benefit from a two-year tax holiday, followed by three years with taxes reduced by half.

19 See Esta Chappell, "Suzhou Explorations: Dushu Lake Higher Education Town," eChinacities.com (January 14, 2013) (http://www.echinacities.com/Suzhou/city-guide/Suzhou-Explorations-Dushu-Lake-Higher-Education-Town).

20 The research was conducted cooperatively between the Institute of Production Science at Karlsruhe Institute of Technology, the Institute for Learning and Innovation in Networks at Karlsruhe University of Applied Sciences, and the Global Advanced Manufacturing Institute in Suzhou.

21 The firms were Beaver Nanotechnologies Co., Ltd, Polynova Technology (subsidiary of Polynova Materials), Suzhou Ltd, Hanano Materials Co. Ltd, Suzhou, and Nano-Micro Technology Company.

22 "In all four cases, the founders (and many co-founders) had taken advanced degrees and/or worked at skilled jobs abroad, specifically in the US and Japan" (Jin et al. 2015: 163).

23 In what SIP terms "promoting the tertiary industry." SIP, "Learn from the World" website (2016).

24 The rest are contract research organizations (CROs), firms engaged in the development of nanomaterials, and investment and finance service providers. See the bioBay website, http://en.biobay.com.cn/main/index.asp.

25 As of September 2013, SIP claimed to have 20 nanotech facilities, 200 nanotech companies "in upstream sectors (nano materials, dispersion technology, and packing), middle stream of core components, and downstream applications)," 200 top talents, and "ties with two dozen industrial communities in the US, Finland, Germany, Singapore, Japan, and UK" (SIP 2013): "Suzhou Industrial Park Innovates Nanotech Development Pattern: An Overview of No. 1 Industry in SIP;" September 29, 2013).

26 Between May 2010 and September 2012, we interviewed three dozen individuals associated with 21 different organizations in Suzhou Industrial Park, Shanghai, and Taixing City (a county-level city in Jiangsu province). Interviews were carried out with principal investigators (PIs) from Chinese research universities and government labs, small startup founders and key entrepreneurs, the central government, and public officials from Shanghai and Suzhou. They included the Shanghai Nanotechnology Promotion Center, the American Chamber of Commerce in Shanghai, and with project managers and senior researchers for SIP's Nanopolis and Suzhou Institute of Nanotech and Nanobionics (SINANO). Interviews focused on two broad themes: respondents' knowledge and views of current government-led nanotech investment strategies and policies, and their perceptions concerning cultural differences in innovation and entrepreneurship between China and the US. More specific questions related to their sources of funding, publication and patent trends, attitudes toward technology transfer, Chinese and international collaboration, training-related background, overseas experiences, and the nature of their research and commercialization activities.

Chapter 6: Xi Jinping's Chinese Dream: Some Challenges

1 To be considered local, however, does not mean living, working, or going to school in (for example) Beijing. Local means having a *hukou*, or residency, in the relevant city.

2 Invitations to participate in the online survey were sent to over 18,000 individuals. A total of 731 individuals completed the survey. The narrative responses below are from the survey, or follow-up face-to-face interviews conducted in China in May 2016.

3 The US, for example, with numerous government- and corporate-sponsored programs, is regarded as leading in training nanotechnology scientists and engineers; more than 75 US colleges and universities currently offer nanotech-related degrees (US NAS 2016: 14–15).

4 In 2009, Schneider settled with Chint Group Corp. for 157.5 million yuan (approximately US$20.7 million), about half of the awarded amount.

5 The OECD average ratio in 2016 was 219 percent; the US ratio was 204 percent (OECD 2017a).

6 Central government policies generally forbid local governments from borrowing from financial markets to fund infrastructure and other development projects. Yet localities are under considerable pressure from Beijing to engage in development-related projects, since performance evaluations heavily favor GDP growth (Anderlini 2013). Local governments have circumvented this restriction by guaranteeing the loans that companies often require to do government work. To take one recent example of the problems this can create, in August 2017 the government of Ningxiang County, Hunan province, invalidated all the guarantees it had issued since 2015. Although the legality of this move is in question, many finance companies and banks are reportedly seeking early loan repayments (Chinascope 2017).

7 See, e.g., Anderlini (2013), Pettis (2013), OECD (2013), Rothman (2015), Shambaugh (2015), and Nederveen Pieterse's (2015) introduction to the special issue of *Third World Quarterly* (36:11, 2015), which is dedicated to this question.

8 The OECD defines the boundary between middle and high income as a per capita GDP of $12,500, at 2011 prices.

9 The survey excluded Tibet, Xinjiang, Hong Kong, and Macao.

10 In August 2017, Lu – China's most dedicated and prominent researcher and blogger of disturbances – was sentenced to four years in prison for "picking quarrels and provoking trouble" (Ramzy 2017).

11 The Plan called for a slight lowering in coal production by 2017, and an overall reduction of $PM_{2.5}$ particles (those with a diameter of less than 2.5 μm) by 15–25 percent, depending on the region. Thousands of factories were closed, fuel and emission standards raised, and many old cars were scrapped (Ouyang 2017).

12 The World Health Organization regards anything above 25 μm as unsafe (Ouyang 2017).

13 The gap between Chinese and US spending on clean energy will likely increase substantially: in January 2017, China's National Energy Administration announced that it planned to spend in excess of $360 billion on solar, wind, and other renewable power sources through 2020, in a program that would create 13 million jobs – at a time when the Trump administration is moving away from renewables in favor of coal and oil (Forsythe 2017).

14 The study, which was conducted under the auspices of the Bank of Finland's Institute for Economies in Transition, drew on data from the Chinese Household Income Project and the China Family Panel Studies. It was a collaborative project that included Beijing Normal University's School of Economics and Business Administration, the Chinese Academy of Social Science's Institute of Economics, with assistance from China's National Bureau of Statistics.

15 A Gini coefficient of 0 indicates perfect equality (everyone has identical income); a coefficient of 1 indicates perfect inequality (one person has all the income, everyone else has none). For purposes of comparison, the Gini coefficient for the US in 2012 was 0.41; France, 0.33; Germany, 0.30; UK, 0.33. On the other hand, based on World Bank data, the corresponding figures for Brazil (0.53), Russia (0.42), India (0.35), and South Africa (0.63) would place China in mid-range (Kanbur et al. 2017: footnote 6).

16 Other income components that were used to determine the degree of inequality included operational income (from farms and private enterprises), property income (rental or sales of property), government transfers, and miscellaneous other income (money and gifts from relatives and friends).

17 As of June 2017, China accounted for 18 percent of a total of $6.2 trillion in US treasuries held by all foreign holders (US FRB 2017a). Less than two decades ago (June 2000), Japan accounted for the lion's share of foreign-held US treasury securities (30 percent), with China only a small fraction (6 percent). China surpassed Japan in 2009 (27 percent v. 21 percent), and reached its high point in 2011 (28 percent), at a time when the US was issuing treasuries as part of its post-recession stimulus efforts (US FRB 2017b). As of July 2017, China's foreign exchange reserves totaled $3.1 trillion, down from a high of $4.0 trillion in June 2014 (Trading Economics 2017b).

18 The IMF approved the Chinese yuan to become a global reserve currency in 2015 alongside the dollar, the euro, the yen, and the British pound (Reuters 2016).

19 In the view of Henderson et al. (2013: 1229), "the Chinese economy may perhaps be most appropriately conceptualized, at this point in time, as a version of 'market neo-Leninism.'"

References

Albert, Eleanor and Beina Xu. 2016. "China's Environmental Crisis," Council on Foreign Relations (January 18). Available at https://www.cfr.org/backgrounder/chinas-environmental-crisis

Amin, Ash and Nigel Thrift. 1992. "Neo-Marshallian Nodes in Global Networks," *International Journal of Urban and Regional Research* 16(4): 571–87.

Anderlini, Jamil. 2013. "The China Road to Reform," *Financial Times* (November 13). Available at https://www.ft.com/content/9bfbdad0-4c6d-11e3-923d-00144feabdc0

Appelbaum, Richard P., Rachel A. Parker, and Cong Cao. 2011a. "Developmental State and Innovation: Nanotechnology in China," *Global Networks* 11(3): 298–314.

Appelbaum, Richard P., Rachel A. Parker, Cong Cao, and Gary Gereffi. 2011b. "China's (Not So Hidden) Developmental State: Becoming a Leading Nanotechnology Innovator in the Twenty-first Century," in *State of Innovation: The US Government's Role in Technology Development*, edited by Fred Block and Matthew R. Keller, 217–35. Boulder, CO: Paradigm Publishers.

Appelbaum, Richard, Cong Cao, Rachel Parker, and Yasuyuki Motoyama. 2012. "Nanotechnology as Industrial Policy: China and the United States," in *The Social Life of Nanotechnology*, edited by Barbara Herr Harthorn and John Mohr, 111–33. New York: Routledge.

Appelbaum, Richard P., Matthew A. Gebbie, Xueying Han, Galen Stocking, and Luciano Kay. 2016. "Will China's Quest for Indigenous Innovation Succeed? Some Lessons from Nanotechnology," *Technology in Society* 46: 149–63.

Aron, Jacob. 2016. "China Launches World's First Quantum Communications Satellite," *New Scientist* (August 16). Available at https://www.newscientist.com/article/2101071-china-launches-worlds-first-quantum-communications-satellite/

Arrow, Kenneth. 1962. "Economic Welfare and the Allocation of Resources for Invention," in *The Rate and Direction of Inventive Activity: Economic and Social*

Factors, edited by Universities-National Bureau Committee for Economic Research, Committee on Economic Growth of the Social Science Research Council, 609-25. Princeton, NJ: Princeton University Press.

Associated Press. 2016. "China's Xi Jinping to Stress Anti-Corruption Campaign Isn't Over," NBC News (October 24). Available at https://www.nbcnews.com/news/china/china-s-xi-jinping-stress-anti-corruption-campaign-isn-t-n671611

Atkinson, Robert D., Nigel Cory, and Stephen J. Ezell. 2017. Stopping China's Mercantilism: A Doctrine of Constructive, Alliance-Backed Confrontation, Information Technology and Innovation Foundation (ITIF) (March). Available at https://itif.org/publications/2017/03/16/stopping-chinas-mercantilism-doctrine-constructive-alliance-backed

Audretsch, David B. and Maryann P. Feldman. 2004. "Knowledge Spillovers and the Geography of Innovation," *Handbook of Regional Urban Economics* 4: 2713-39.

Baark, Erik. 2014. Is China Becoming a Science and Technology Superpower, and So What? Discussion paper presented at conference entitled "The Evolving Role of Science and Technology in China's International Relations," April 3-4, 2014, Arizona State University, Tempe, AZ.

Bai, Chunli. 2017a. "International S&T Cooperation Network Will Be Established by 2030." National Natural Science Foundation of China. Available at http://www.nsfc.gov.cn/publish/portal0/tab434/info68550.htm

Bai, Chunli. 2017b. ORI Network of International S&T Cooperation to Be Established by 2030. Xinhua. Available at http://news.xinhuanet.com/tech/2017-05/10/c_1120945651.htm

Baidu. 2017. "Robin Li." Available at: http://ir.baidu.com/phoenix.zhtml?c=188488&p=irol-govBio&ID=138201

Baidu. 2018. "Autonomous Driving Unit." Available at: http://usa.baidu.com/adu/

Baker, David R. 2017. "China's Baidu Opens Self-Driving Lab in Silicon Valley." *SFGate* (October 4). Available at http://www.sfgate.com/news/article/China-s-Baidu-opens-self-driving-lab-in-Silicon-12253223.php

Barboza, David. 2006. "In a Scientist's Fall, China Feels Robbed of Glory," *The New York Times* (May 15). Available at http://www.nytimes.com/2006/05/15/technology/15fraud.html

Batjargal, Bat and Manhong Liu. 2004. "Entrepreneurs' Access to Private Equity in China: The Role of Social Capital," *Organization Science* 15(2): 159-72.

BBC News. 2012. "Full Text: China's New Party Chief Xi Jinping's Speech" (November 5). Available at http://www.bbc.co.uk/news/world-asia-china-20338586

BBC News. 2016. "China Launches Quantum-Enabled Satellite Micius" (August 16). Available at http://www.bbc.com/news/world-asia-china-37091833

BCG. 2017a. "Innovation in 2016," bcg.perspectives (January 12). Available at https://www.bcg.com/publications/2017/growth-innovation-in-2016.aspx

BCG. 2017b. "BCG Survey Names 2016's 50 Most Innovative Companies," BCG press release (January 12). Available at https://www.bcg.com/d/press/12january2017-most-innovative-companies-2016-142287

BCG. 2017c. "The Most Innovative Companies: An Interactive Guide," bcg.perspectives (January 12). Available at https://www.bcgperspectives.com/content/interactive/innovation_growth_most_innovative_companies_interactive_guide/

Beaver, Laurie. 2016. "WeChat Breaks 700 Million Monthly Active Users." *Business Insider* (April 20). Available at http://www.businessinsider.com/wechat-breaks-700-million-monthly-active-users-2016-4

Bhagwati, Jagdish N. (ed.). 1977. *The New International Economic Order: The North-South Debate*. Cambridge, MA: MIT Press.

Bornmann, Lutz, Caroline S. Wagner, and Loet Leydesdorff. 2015. "BRICS Countries and Scientific Excellence: A Bibliometric Analysis of Most Frequently Cited Papers," *Journal of the Association for Information Science and Technology* 66(7): 1507–13.

Boschma, Ron. 2005. "Proximity and Innovation: A Critical Assessment," *Regional Studies* 39(1): 61–74.

Bound, Kristen, Tom Saunders, James Wilsdon, and Jonathan Adams. 2013. *China's Absorptive State: Research, Innovation, and the Prospects for China-UK Collaboration*. London: NESTA.

Breznitz, Dan and Michael Murphree. 2011. *Run of the Red Queen: Government, Innovation, Globalization, and Economic Growth in China*. New Haven, CT: Yale University Press.

Bunnel, Tom. 2004. *Malaysia, Modernity and the Multimedia Super Corridor*. London: RoutledgeCurzon.

Burt, Ronald. 1992. *Structural Holes: The Social Structure of Competition*. Cambridge, MA: Harvard University Press.

Business Standard. 2016. "Beijing to Upgrade Local 'Silicon Valley'" (April 11). Available at http://www.business-standard.com/article/news-ians/beijing-to-upgrade-local-silicon-valley-116041100506_1.html.

Cadell, Cate. 2017. "China's Baidu Launches $1.5 Billion Autonomous Driving Fund." Reuters (September 21). Available at https://www.reuters.com/article/us-china-baidu-autonomous/chinas-baidu-launches-1-5-billion-autonomous-driving-fund-idUSKCN1BW0QJ

Cao, Cong. 2002. "Strengthening China through Science and Education: China's New Development Strategy toward the Twenty-First Century," *Issues & Studies* 38(3): 122–49.

Cao, Cong. 2004. "Zhongguancun and China's High-Tech Parks in Transition: 'Growing Pains' or 'Premature Senility'?" *Asian Survey* 44(5): 647–68.

Cao, Cong. 2013. "Science Imperiled in the Cultural Revolution," in *Mr. Science and Chairman Mao's Cultural Revolution: Science and Technology in Modern China*, edited by Chunjuan Nancy Wei and Darryl E. Brock, 119–42. Lanham, MD: Lexington Books.

Cao, Cong and Richard P. Suttmeier. 2017. "Challenges of S&T System Reform in China," *Science* 335: 1019–21.

Cao, Cong, Richard P. Suttmeier, and Denis Fred Simon. 2006. "China's 15-Year Science and Technology Plan," *Physics Today* 59(12): 38–43.

Cao, Cong, Richard P. Suttmeier, and Denis Fred Simon. 2009. "Success in State Directed Innovation? Perspectives on China's Plan for the Development of Science and Technology," in *The New Asian Innovation Dynamics: China and India in Perspective*, edited by Govindan Parayil and Anthony P. D'Costa, 247–64. London: Palgrave Macmillan.

Cao, Cong, Ning Li, Xia Li, and Li Liu. 2013a. "Reforming China's S&T System," *Science* 341: 460–2.

Cao, Cong, Richard Appelbaum, and Rachel Parker. 2013b. "Research is High and the Market is Far Away – Commercialization of Nanotechnology in China," *Technology in Society* 35: 55–64.

Cao, Qing. 2014. "Insight into Weak Enforcement of Intellectual Property Rights in China," *Technology in Society* 38: 40–7.

CAS. 2017. "Report on State of One Belt One Road S&T Cooperation" (May 9). Chinese Academy of Sciences and Elsevier. Available at http://www.cas.cn/sygz/201705/t20170509_4600090.shtml

CCP's Central Committee and State Council. 2016. *Outline of the National Innovation-Driven Development Strategy*. Available at http://politics.people.com.cn/n1/2016/0520/c1001-28364670.html (in Chinese).

CEIC. 2017. "China Non Performing Loans Ratio," CEICV: A Euromoney Institutional Investor Company. Available at https://www.ceicdata.com/en/indicator/china/non-performing-loans-ratio

Chen, Shanzhi, Hui Xu, Dake Liu, Bo Hu, and Hucheng Wang. 2014. "A Vision of IoT: Applications, Challenges, and Opportunities with China Perspective," *IEEE Internet of Things Journal* 1(4): 349–59.

Chen, Te-Ping. 2016. "Test Driven: Parents Protest Changes to University Entrance Exam," *The Wall Street Journal* (May 15). Available at https://blogs.wsj.com/chinarealtime/2016/05/15/test-driven-parents-protest-changes-to-university-entrance-exam/

Cheng, Ruyan. 2008. "China's International Science and Technology Cooperation Strategy and Policy Evolution in 30 Years," *China S&T Forum* 7(26): 7–11 (in Chinese).

China Daily. 2014. "Top Tech: Zhongguancun Science Park." Available at http://www.chinadaily.com.cn/beijing/2014-05/16/content_17511673.htm

China Daily. 2017a. "Zhongguancun, Cluster of Unicorn Companies." Available at http://www.chinadaily.com.cn/m/beijing/zhongguancun/2017-03/03/content_28418152.htm

China Daily. 2017b. "China Pushes for Global Internet of Things Standard." Available at http://www.chinadaily.com.cn/business/tech/2017-10/20/content_33495833.htm

China Daily. 2017c. "China's 'Smart Cities' to Number 500 before end of 2017." Available at http://www.chinadaily.com.cn/china/2017-04/21/content_29024793.htm

China News. 2018. "Guangdong Ranked First in China's Overall Capability of Innovation and Innovation" (January 25). Available at http://news.163.com/18/0125/14/D90JB88D00018AOQ.html

China State Council. 2015. "State Council Approves Plan for Industrial Park in Suzhou" (October 13). Available at http://english.gov.cn/policies/latest_releases/2015/10/13/content_281475210782608.htm

China Statistical Yearbook. 2006. "Table 21-9 Number of Postgraduate Students by Field of Study (2005)." Available at http://www.stats.gov.cn/tjsj/ndsj/2006/indexeh.htm

China Statistical Yearbook. 2016. "Table 21-11 Number of Postgraduate Students by Academic Field (2015)." Available at http://www.stats.gov.cn/tjsj/ndsj/2016/indexeh.htm

China Statistical Yearbook. 2017. Table 18, "Reference Exchange Rate of Renminbi (Period Average)." Available at http://www.stats.gov.cn/tjsj/ndsj/2017/indexeh.htm

China STI. 2016. "The 13th Five-year National Plan for Science, Technology and Innovation of the People's Republic of China" (August).

China STI. 2017. 13th Five-Year Plan on STI International Cooperation Promulgated, *China Science and Technology Newsletter*, no. 11 (June 15).

Chinano 2018. "Chinano Conference and Expo," October 24–26, Suzhou, China. Available at https://10times.com/chinano-conference-expo

Chinascope. 2017. "A Hunan Government Invalidated All Its Government Loan Guarantees Overnight" (August 28). Available at http://chinascope.org/archives/13082

Chinese Academy of Sciences. 2016. "Line for Quantum Communication to be Ready Next Year." Available at http://english.cas.cn/newsroom/news/201611/t20161124_171025.shtml

Cloud Security Alliance. 2017. "Quantum-Safe Security Working Group." Available at https://cloudsecurityalliance.org/group/quantum-safe-security/

Colson, Thomas. 2016. "The Top 10 Countries Where Chinese Students Study Abroad." *Business Insider* (September 16). Available at http://www.businessinsider.com/knight-frank-ranking-countries-where-chinese-students-study-abroad-2016-9/#10-new-zealand-13952-students-1

Covington. 2015. "SIPO is Reinforcing Its Role in Patent Protection in China." Available at https://www.cov.com/-/media/files/corporate/publications/2015/07/sipo_is_reinforcing_its_role_in_patent_protection_in_china

CSSD. 2014. *Eco-Environment Protection and Introduction to China Singapore Eco-Technology Park*. China-Singapore Suzhou Industrial Park Development Co., Ltd. (CSSD) PowerPoint presentation. http://www.siww.com.sg/pdf/forum/CBF1-ProtectionAndIntroductionToEcoTechPark.pdf.

Custer, Charlie. 2016. "3 Reasons Why WeChat Failed Internationally (and most other Chinese apps do too)," *TechinAsia* (May 25). Available at https://www.techinasia.com/3-reasons-wechat-failed-internationally-chinese-apps

Cuthbertson, Anthony. 2017. "China's Baidu Aims to Challenge Google Maps' Dominance." *Newsweek* (January 17). Available at: http://www.newsweek.com/google-maps-baidu-map-here-china-543682

Dolven, Ben. 1999. "Suzhou Project: Wounded Pride," *Far Eastern Economic Review* (July 8). Available at http://web.archive.org/web/20060512193509/http://www.sfdonline.org/Link%20Pages/Link%20Folders/Other/suzhou3.html

Dong, Liansai. 2015. "367 Shades of Grey: Why China Needs a Coal Cap," Greenpeace East Asia (October 15). Available at http://www.greenpeace.org/eastasia/news/blog/367-shades-of-grey-why-china-needs-a-coal-cap/blog/54429/

Doud, Adam. 2017. "What's Keeping Xiaomi from the US?" *Android Authority* (October 17). Available at https://www.androidauthority.com/xiaomi-us-803588/

Duncan, Jeffery M., Michelle A. Sherwood, and Yuanlin Shen. 2008. "A Comparison between the Judicial and Administrative Routes to Enforce Intellectual Property Rights in China," *The John Marshall Review of Intellectual Property Law* 7(3): 529–44.

Dutta, Soumitra, Rafael Escalona Reynoso, Jordan Litner, Bruno Lanvin, Sacha Wunsch-Vincent, and Kritika Saxena. 2016. "The Global Innovation Index 2016: Winning with Global Innovation," in *The Global Innovation Index 2016*, edited by Soumitra Dutta, Bruno Lanvin, and Sacha Wunsch-Vincent, 3–74.

Geneva: WIPO. Available at www.wipo.int/edocs/pubdocs/en/wipo_pub_gii_2016-chapter1.pdf

Eichengreen, Barry, Donghyun Park, and Kwanho Shin. 2013. "Growth Slow-downs Redux: New Evidence of the Middle-Income Trap," National Bureau of Economic Research Working Paper 18673. Available at http://www.nber.org/papers/w18673.pdf

Embassy of PRC in Japan. 2007. "General Situations of Sino-Japan Science and Technology Cooperation." *China Daily.*

Embassy of PRC in Japan. 2016. Japan Science and Technology Agency and Ministry of Science and Technology China Initiate Joint R&D Projects. Available at http://www.fmprc.gov.cn/ce/cejp/jpn/sgxw/t1408125.htm

Etzkowitz, Henry and Loet Leydesdorff. 2000. "The Dynamics of Innovation: From National Systems and 'Mode 2' to a Triple Helix of University–Industry–Government Relations," *Research Policy* 29(2), 109–23.

Euratom. 2008. "Agreement between the European Atomic Energy Community and the Government of the People's Republic of China for R&D Cooperation in the Peaceful Uses of Nuclear Energy" (April 24). Available at https://eeas.europa.eu/sites/eeas/files/r_d-pune_agreement_en.pdf

European Commission. 2002. *Benchmarking of Business Incubators: Final Report.* European Commission Enterprise Directorate-General (February).

European Commission. 2018. *Report on the Protection and Enforcement of Intellectual Property Rights in Third Countries.* Brussels: European Commission.

EU. 2015. "EU–China Research and Innovation Relations." Available at http://ec.europa.eu/research/iscp/pdf/policy/1_eu_CHINA_R_I.pdf#view=fit&pagemode=none

European Union Chamber of Commerce in China. 2017. "China Manufacturing 2025: Putting Industrial Policy Ahead of Market Forces." Available at http://docs.dpaq.de/12007-european_chamber_cm2025-en.pdf

export.gov. 2017. "China Country Commercial Guide – State Owned Enterprises (July 25). Available at https://www.export.gov/article?id=China-State-Owned-Enterprises.

Fan, Wei, Yun Liu, Luciano Kay, and Yije Cheng. 2014. "Two Poles in Global Nano Research: Structure and Evolution of the Global Nano Collaborative Innovation Network," Proceedings of PICMET 14 Conference: Portland International Center for Management of Engineering and Technology; Infrastructure and Service Integration.

Fang, Jing, Hui He, and Nan Li. 2016. "China's Rising IQ (Innovation Quotient) and Growth: Firm-Level Evidence." IMF Working Paper WP/16/249. Available at https://www.imf.org/external/pubs/ft/wp/2016/wp16249.pdf

Feigenbaum, Evan A. 2003. *China's Techno-Warriors: National Security and Strategic Competition from the Nuclear to the Information Age.* Stanford, CA: Stanford University Press.

Fiegerman, Seth. 2017. "Facebook Tops 1.9 Billion Monthly Users." *CNN Tech* (May 3). Available at http://money.cnn.com/2017/05/03/technology/facebook-earnings/index.html

Forbes. 2017. "Forbes 2017 Billionaires List: Mainland Chinese Make Up the Greatest Number of New Entrants," *Forbes* (March 21). Available at http://www.channelnewsasia.com/news/business/forbes-2017-billionaires-list-mainland-chinese-make-up-greatest—8582612

Forsythe, Michael. 2017. "China Cancels 103 Coal Plants, Mindful of Smog and Wasted Capacity," *New York Times* (January 18). Available at: https://www.nytimes.com/2017/01/18/world/asia/china-coal-power-plants-pollution.html

Fortune Magazine. 2016. "The Unicorn List: 2016." Available at: http://fortune.com/unicorns/.

Fortune Magazine. 2018. "The Global 500 list: 2017." Available at http://fortune.com/global500/list/filtered?hqcountry=China

Fu, Yiqin. 2013. "China's Unfair College Admissions System." *The Atlantic* (June 19). Available at https://www.theatlantic.com/china/archive/2013/06/chinas-unfair-college-admissions-system/276995/

Galbraith, Kate. 2014. "My Name is China, and I Have a Pollution Problem," *Foreign Policy* (May 29). Available at http://foreignpolicy.com/2014/05/29/my-name-is-china-and-i-have-a-pollution-problem/

Gartner. 2017. "Gartner Says 8.4 Billion Connected 'Things' Will Be in Use in 2017, Up 31 Percent from 2016." Available at: https://www.gartner.com/newsroom/id/3598917

Ghafele, Roya and Benjamin Gibert. 2012. "Promoting Intellectual Property Monetization in Developing Countries: A Review of Issues and Strategies to Support Knowledge-Driven Growth," Policy Research Working Series 6143, World Bank. Available at http://elibrary.worldbank.org/doi/pdf/10.1596/1813-9450-6143

Gill, Indermit S. and Homi Kharas. 2007. *An East Asian Renaissance: Ideas for Economic Growth.* Washington, DC: World Bank.

Glänzel, Wolfgang, Koenraad Debackere, and Martin Meyer. 2008. "'Triad' or 'Tetrad'? On Global Changes in a Dynamic World," *Scientometrics* 74(1): 71–88.

Gong, Yidong. 2005. "China Science Foundation Takes Action against 60 Grantees," *Science* 309(5742): 1798–9.

Gray, Richard. 2017. "China's Zombie Factories and Unborn Cities," BBC Future (February 23). Available at http://www.bbc.com/future/story/20170223-chinas-zombie-factories-and-unborn-cities

Greene, J. Megan. 2008. *The Origins of the Developmental State in Taiwan: Science Policy and the Quest for Modernization.* Cambridge, MA: Harvard University Press.

Griffiths, James. 2016. "Torture in Secret Prisons: The Dark Side of China's Anti-Corruption Campaign," CNN World (December 6). Available at http://www.cnn.com/2016/12/06/asia/china-shuanggui-communist-party-torture/index.html

Grimes, Seamus and Marcela Miozzo. 2015. "Big Pharma's Internationalization of R&D to China," *European Planning Studies* 23(9): 1873–94.

Guerrini, Federico. 2016. "How Beijing Is Using Data from Social Media and Iot to Boost Air Pollution Forecasting," *Forbes* (May 21). Available at https://www.forbes.com/sites/federicoguerrini/2016/05/21/how-beijing-is-using-big-data-from-social-media-and-iot-to-improve-air-pollution-management/#5a09ab225c80

Hackett, Robert. 2017. "IBM Sets Sight on Quantum Computing," *Fortune* (March 5). Available at http://fortune.com/2017/03/06/ibm-quantum-computer/

Haltmaier, Jane. 2013. "Challenges for the Future of Chinese Economic Growth," US Federal Reserve Board of Governors, International Finance Discussion Paper 1072.

Han, Xueying and Richard P. Appelbaum. 2016. Will They Stay or Will They Go? International STEM Students Are Up for Grabs. Kansas City, MO: Ewing Marion Kauffman Foundation. Available at http://www.kauffman.org/~/media/kauffman_org/research%20reports%20and%20covers/2016/stem_students_final.pdf

Han, Xueying and Richard P. Appelbaum. 2018. "China's Science, Technology, Engineering and Mathematics (STEM) Research Environment: A Snapshot," *PLOS One* (forthcoming).

He, Huifeng, 2017. "Premier Li Keqiang's Innovation Push Proves No Miracle Cure for China's Economy," *South China Morning Post* (March 9). Available at http://www.scmp.com/news/china/policies-politics/article/2076391/young-take-lead-chinese-premiers-innovation-push

Heilmann, Sebastian, Lea Shih, and Andreas Hofman. 2013. "National Planning and Local Technology Zones: Experimental Governance in China's Torch Programme," *The China Quarterly* 216: 896–919.

Henderson, Jeffrey, Richard P. Appelbaum, and Suet Ying Ho. 2013. "Globalization with Chinese Characteristics: Externalization, Dynamics and Transformation," *Development and Change* 44(6): 1221–53.

HKTDC. 2015. "China–Singapore Suzhou Industrial Park (including Suzhou Industrial Park Export Processing Zone)." HKTDC Research (May 7). Available at http://china-trade-research.hktdc.com/business-news/article/Fast-Facts/China-Singapore-Suzhou-Industrial-Park-including-Suzhou-Industrial-Park-Export-Processing-Zone/ff/en/1/1X000000/1X09WGSW.htm

Hou, Qiang. 2013. "The Conclusion and Implementation of the Sino-Soviet S&T Cooperation Agreement in Early PRC," *Siberian Studies* 40(5).

Howells, Jeremy R.L. 2002. "Tacit Knowledge, Innovation and Economic Geography," *Urban Studies* 39(5–6): 871–84.

Hsu, Sara. 2017. "Foreign Firms Wary of 'Made in China 2025,' but it May Be China's Best Chance at Innovation," *Forbes* (March 10). Available at https://www.forbes.com/sites/sarahsu/2017/03/10/foreign-firms-wary-of-made-in-china-2025-but-it-may-be-chinas-best-chance-at-innovation/#2a17140624d2

Hu, Xiaoyu and Wei Lu. 2017. "Wuxi Cements Position as National Internet of Things Center," *China Daily* (September 9). Available at http://www.chinadaily.com.cn/cndy/2017-09/09/content_31757712.htm

Huang, Yasheng. 2008. *Capitalism with Chinese Characteristics: Entrepreneurship and the State*. New York: Cambridge University Press.

Huawei. 2016. "Huawei and China Unicom (Shanghai) Jointly Release the NB-IoT-based Smart Parking Solution" (July 1). Available at http://www.huawei.com/en/news/2016/7/NB-IoT-based-Smart-Parking-Solution

Hubbard, Paul, Samuel Hurley, and Dhruv Sharma. 2012. "The Familiar Pattern of Chinese Consumption Growth," *Economic Roundup*, Issue 4. Available at https://treasury.gov.au/publication/economic-roundup-issue-4-2012/the-familiar-pattern-of-chinese-consumption-growth/

Human Rights Watch. 2017. "China: Voice Biometric Collection Threatens Privacy." Available at: https://www.hrw.org/news/2017/10/22/china-voice-biometric-collection-threatens-privacy

Hvistendahl, Mara. 2013. "China's Publication Bazaar," *Science* 342(6162): 1035–9.

Hvistendahl, Mara. 2015. "China Pursues Fraudsters in Science Publishing," *Science* 350(6264): 1015.

IASP. 2015. "Science Park (IASP Official Definition)." International Association of Science Parks and Areas of Innovation. Available at http://www.iasp.ws/Our-industry/Definitions

IDG China. 2015. "Haidian Park: The Birthplace of China's Most Innovative and Entrepreneurial Technology Companies," CISION PR Newswire, Chinese Silicon Valley Series (January 22). Available at http://www.prnewswire.com/news-releases/haidian-park-the-birthplace-of-chinas-most-innovative-and-entrepreneurial-technology-companies-300024231.html

IEEE Smart Cities. 2017. "IEEE Core Smart City Profile – Wuxi, China." Available at https://smartcities.ieee.org/core-cities/wuxi-china.html

IMF. 2014. "Table 5: Report for Selective Countries and Subjects," *World Economic Outlook 2014*. Database available at http://www.imf.org/external/pubs/ft/weo/2014/02/weodata/index.aspx

IMF. 2017. "IMF DataMapper: Real GDP Growth, Annual Percent Change." International Monetary Fund, World Economic Outlook (October). Available at http://www.imf.org/external/datamapper/NGDP_RPCH@WEO/OEMDC/ADVEC/WEOWORLD/CHN

Indo Asian News Service. 2017. "Xiaomi Shares Top Slot with Samsung in India in Q3, 2017: IDC," *Hindustantimes* (November 14). Available at http://www.hindustantimes.com/tech/xiaomi-shares-top-slot-with-samsung-in-india-in-q3-2017-idc/story-tiD5fpptqQFANI7rSf1HQI.html

Institute of International Education (IIE). 2010. "Top 25 Places of Origin of International Students, 2008/09–2009/10." *Open Doors Report on International Educational Exchange*. Available at http://www.iie.org/opendoors

Institute of International Education. 2011. "Top 25 Places of Origin of International Students, 2000/10–2010/11." *Open Doors Report on International Educational Exchange*. Available at http://www.iie.org/opendoors

Institute of International Education. 2012. "Top 25 Places of Origin of International Students, 2010/11–2011/12." *Open Doors Report on International Educational Exchange*. Available at http://www.iie.org/opendoors

Institute of International Education. 2013. "Top 25 Places of Origin of International Students, 2011/12–2012/13." *Open Doors Report on International Educational Exchange*. Available at http://www.iie.org/opendoors

Institute of International Education. 2014. "Top 25 Places of Origin of International Students, 2012/13–2013/14." *Open Doors Report on International Educational Exchange*. Available at http://www.iie.org/opendoors

Institute of International Education. 2015. "Top 25 Places of Origin of International Students, 2013/14–2014/15." *Open Doors Report on International Educational Exchange*. Available at http://www.iie.org/opendoors

Institute of International Education. 2016. "Top 25 Places of Origin of International Students, 2014/15–2015/16." *Open Doors Report on International Educational Exchange*. Available at http://www.iie.org/opendoors

International Federation of Robotics. 2016. "Executive Summary World Robotics 2016 Industrial Robots." Available at https://ifr.org/img/uploads/Executive_Summary_WR_Industrial_Robots_20161.pdf

Jaeger, Joel, Paul Joffe, and Ranping Song. 2017. "China is Leaving the US Behind on Clean Energy Investment," World Resources Institute

(January 6). Available at http://www.wri.org/blog/2017/01/china-leaving-us-behind-clean-energy-investment

Jersild, Austin, 2014. *The Sino-Soviet Alliance: An International History*. Chapel Hill, NC: University of North Carolina Press.

Jiang, Sijia. 2017. "Tencent Steps Up AI Push with Research Lab in Seattle," *Technology News* (May 2). Available at https://uk.reuters.com/article/us-tencent-ai/tencent-steps-up-ai-push-with-research-lab-in-seattle-idUKKBN17Y0EU

Jin, Jun, Zhengyi Zhang, and Maureen McKelvey. 2015. "The Emergence of Knowledge Intensive Entrepreneurship in China: Four Start-up Companies in Nanotechnology in Suzhou," in *Innovation Spaces in Asia: Entrepreneurs, Multinational Enterprises and Policy*, edited by Maureen McKelvey and Sharmista Bagchi-Sen. Cheltenham: Edward Elgar.

Johnson, Chalmers. 1982. *MITI and the Japanese Miracle: The Growth of Industrial Policy, 1925–1975*. Stanford, CA: Stanford University Press.

Johnston, Ian. 2017. "Quantum Computing Breakthrough Could Help 'Change Life Completely', Say Scientists." *Independent* (February 1). Available at http://www.independent.co.uk/news/science/quantum-computers-quantum-physics-sussex-university-holy-grail-a7558036.html

JRJ. 2012. "Xu Guanhua: The Transformation of the Chinese Economy Is Facing Three Fundamental Challenges" (in Chinese). Available at http://finance.jrj.com.cn/people/2012/05/31143113329911.shtml

Kanbur, Ravi, Yue Wang, and Xiaobo Zhang. 2017. "The Great Chinese Inequality Turnaround," BOFIT Discussion Papers 6:2017. Bank of Finland, BPOFIT: Institute for Economics in Transition (February 5). Available at https://helda.helsinki.fi/bof/bitstream/123456789/14667/1/dp0617.pdf

Kennedy, Scott. 2015. "Critical Questions: Made in China 2025." Center for Strategic and International Studies (June 1). Available at https://www.csis.org/analysis/made-china-2025

Kharas, Homi. 2017. "The Unprecedented Expansion of the Global Middle Class: An Update," Brookings Institution Global Economy and Development Report, Working Paper 100. Available at https://www.brookings.edu/research/the-unprecedented-expansion-of-the-global-middle-class-2/

Knight, Will. 2015. "China Wants to Replace Millions of Workers with Robots," *MIT Technology Review* (December 7). Available at https://www.technologyreview.com/s/544201/china-wants-to-replace-millions-of-workers-with-robots/

Koleski, Katherine. 2017. "US–China Economic and Security Review Commission: The 13th Five-Year Plan." Available at https://www.uscc.gov/sites/default/files/Research/The%2013th%20Five-Year%20Plan_Final_2.14.17_Updated%20%28002%29.pdf

Koty, Alexander Chipman and Zhou Qian, 2017. "A Complete Guide to 2017 Minimum Wage Levels Across China," *China Briefing* (November 15). Available at http://www.china-briefing.com/news/2017/11/15/complete-guide-2017-minimum-wage-levels-across-china.html

KPMG. 2017. *The Changing Landscape of Disruptive Technologies: Global Technology Innovation Hubs*. KPMG International Cooperative. Available at https://info.kpmg.us/content/dam/info/tech-innovation/disruptive-tech-2017-part1.pdf

Kraemer, Kenneth L., Greg Linden, and Jason Dedrick. 2011. Capturing Value in Global Networks: Apple's iPad and iPhone. Unpublished paper. Available at http://economiadeservicos.com/wp-content/uploads/2017/04/value_ipad_iphone.pdf

Lai, Richard. 2017. "Xiaomi Still Isn't Ready to Sell Phones in America." *Engadget* (April 24). Available at https://www.engadget.com/2017/04/24/xiaomi-svp-wang-xiang-interview/

Lancet. 2017. "Editorial: China's Health Trajectory in 2017," *Lancet* 18(2): 155. Available at http://thelancet.com/journals/lanonc/article/PIIS1470-2045(17)30034-7/fulltext

Lanza, Gisela et al. 2015. *Industrial Synergies between Baden-Wuerttemberg and Suzhou Industrial Park*. Baden-Wuerttemberg: Ministry of Science, Research and the Arts.

Lardy, Nicholas and Nicholas Borst. 2013. "A Blueprint for Rebalancing the Chinese Economy," Peterson Institute for International Economics, Policy Brief No. PB13-02. Available at http://www.iie.com/publications/pb/pb13-2.pdf

Lawton, Jim. 2017. "In the Race to Advance Manufacturing, China Is Betting on Robots," *Forbes* (September 21). Available at https://www.forbes.com/sites/jimlawton/2017/09/21/in-the-race-to-advance-manufacturing-chinas-betting-on-robots/#4ee8fded78cd

Le Corre, Philippe and Alain Sepulchre. 2016. *China's Offensive in Europe*. Washington, DC: Brookings Institution Press.

Lee, Chong-Moon, William F. Miller, Marguerite Gong Hancock, and Henry S. Rowen. 2000. "The Silicon Valley Habitat," in *The Silicon Valley Edge: A Habitat for Innovation and Entrepreneurship*, edited by Chong-Moon Lee, William F. Miller, Marguerite Gong Hancock, and Henry S. Rowen, 1–15. Stanford, CA: Stanford University Press.

Lee, Emily. 2015. "Problems of Judicial Recognition and Enforcement in Cross-Border Insolvency Matters between Hong Kong and Mainland China," *American Journal of Comparative Law* 63(2): 439–65.

Lewis, John W. and Litai Xue. 1991. *China Builds the Bomb*. Stanford, CA: Stanford University Press.

Levy, Nat. 2017. "Baidu Presidents Calls AI the 'Single Most Transformative Force of Our Time," *Geekwire* (October 9). Available at https://www.geekwire. com/2017/baidu-president-calls-ai-single-transformative-force-time-talks-u-s-china-twin-engines-innovation/

Lewin, Tamar. 2012. "Taking More Seats on Campus, Foreigners Also Pay the Freight," *The New York Times* (February 4). Available at http://www. nytimes.com/2012/02/05/education/international-students-pay-top-dollar-at-us-colleges.html

Leydesdorff, Loet. 2013. "The Triple Helix of University–Industry–Government Relations," in *Encyclopedia of Creativity, Innovation, and Entrepreneurship*, edited by Elias Carayannis and David Campbell. New York: Springer.

Li, Cheng (ed.). 2005. *Bridging Minds across the Pacific: US–China Educational Exchange, 1978–2003*. Lanham, MD: Lexington Books.

Li, Haizheng. 2010. "Higher Education in China: Complement or Competition to US Universities?" in *American Universities in a Global Market*, edited by Charles T. Clofelter, 269–304. Chicago, IL: University of Chicago Press.

Li, Keqiang. 2018. "Government Work Report 2018," March 5. Available at http://www.gov.cn/zhuanti/2018lh/2018zfgzbg/mobile.htm

Li, Peng. 2018. "Chinese Patents Enter a New Era of Development: From Quantity to Victory and Quality Improvement," *stdaily* (January 26). Available at http://www.stdaily.com/index/kejixinwen/2018-01/26/content_629287.shtml

Li, Yongling, Yanliu Lin, and Stan Geertman. 2015. "The Development of Smart Cities in China," Proceedings of the 14th International Conference on Computers in Urban Planning and Urban Management, July 7–10, Cambridge, MA, USA.

Lieberman, Joseph. 2005. "Foreword," in *Nanotechnology: Science, Innovation, and Opportunity*, edited by Lynn E. Foster. Upper Saddle River, NJ: Prentice-Hall.

Literature Research Office of the CCP's Central Committee (comp.). 2016. *Excerpts of Xi Jinping's Speeches on Science, Technology and Innovation* (in Chinese). Beijing: Central Literature Publishing House.

Liu, Coco. 2017. "Why Are Middle Class Chinese Moving Their Money Abroad?" *South China Morning Post* (May 27). Available at http://www. scmp.com/week-asia/society/article/2095827/why-are-middle-class-chinese-moving-their-money-abroad

Liu, Feng-chao, Denis Fred Simon, Yu-tao Sun, and Cong Cao. 2011. "China's Innovation Policies: Evolution, Institutional Structure, and Trajectory," *Research Policy* 40(7): 917–31.

Lockett, Hudson. 2016. "China Anti-Corruption Campaign Back-fires," *Financial Times* (October 9). Available at https://www.ft.com/content/02f712b4-8ab8-11e6-8aa5-f79f5696c731

Loudenback, Tanza. 2016. "International Students Are Now 'Subsidizing' Public American Universities to the Tune of $9 Billion a Year," *Business Insider* (September 16). Available at http://www.businessinsider.com/foreign-students-pay-up-to-three-times-as-much-for-tuition-at-us-public-colleges-2016-9

Lu, Qiwen. 2000. *China's Leap into the Information Age: Innovation and Organization in the Computer Industry*. New York: Oxford University Press.

Lucas, Louise. 2018. "Facebook, Amazon Handicapped as they Follow China Playbook," *Financial Times* (March 14). Available at https://www.ft.com/content/a450a7ca-25d4-11e8-b27e-cc62a39d57a0

Lux. 2014. *Nanotechnology Update: Corporations Up Their Spending as Revenue for Nanotechnology Projects Increase*. Lux Research: State of the Market Report (February 17). Available at https://members.luxresearchinc.com/research/report/13748

Markusen, Ann. 1996. "Sticky Places in Slippery Space: A Typology of Industrial Districts," *Economic Geography* 72: 293–313.

Marshall, Alfred. 1920 [1890]. *Principles of Economics*, 8th edn. Book IV, chapter X. London: Macmillan and Company. Available at: http://www.econlib.org/library/Marshall/marP24.html#Bk.IV,Ch.X.

Mazzucato, Mariana. 2013. *The Entrepreneurial State: Debunking Public vs. Private Sector Myths*. London: Anthem Press.

McGregor, James. 2010. *China's Drive for Indigenous Innovation: A Web of Industrial Policies*. US Chamber of Commerce Global Regulatory Cooperation Project. Available at https://www.uschamber.com/report/china%E2%80%99s-drive-indigenous-innovation-web-industrial-policies

McKelvey, Maureen and Astrid Heidemann Lassen. 2013. *Managing Knowledge Intensive Entrepreneurship*. Cheltenham: Edward Elgar.

McPhillips, Deidre. 2016. "Where in the World Chinese Students Are Studying," *US News* (October 25). Available at https://www.usnews.com/news/best-countries/articles/2016-10-25/an-international-education-for-chinese-students-abroad

Mehta, Aashish, Patrick Herron, Yasuyuki Motoyama, Richard Appelbaum, and Timothy Lenoir. 2012. "Globalization and De-globalization in Nanotechnology Research: The Role of China," *Scientometrics* 93(2): 439–58.

Miller, Tom. 2012. *China's Urban Billion*. London: Zed Books.

Ministry of Education. 2016. "Number of Students of Formal Education by Type and Level," Educational Statistics in 2015. Available at http://en.moe.gov.cn/Resources/Statistics/edu_stat_2015/2015_en01/201610/t20161012_284510.html

Ministry of Science and Technology. 2006. "National Medium and Long Term Science and Technology Development Plan (2006–2020)" (in Chinese). Available at http://www.most.gov.cn/ztzl/gjzcqgy/zcqgygynr/1.htm

Ministry of Science and Technology. 2014. "China Science and Technology Indicators, 2014." The Yellow Book on Science and Technology, Vol. 12. Available at 2015.casted.org.cn/upload/news/Attach-20161226101958.pdf

Ministry of Science and Technology. 2016. "The Thirteenth Five-Year Plan for Science, Technology, and Innovation" (in Chinese). Available at http://www.most.gov.cn/mostinfo/xinxifenlei/gjkjgh/201608/t20160810_127174.htm

Moody, Andrew. 2015. "High-Tech Zones Up the Game," China Daily (October 9). Available at http://usa.chinadaily.com.cn/epaper/2015-10/09/content_22140765.htm

Motoyama, Yasuyuki, Cong Cao, and Richard Appelbaum. 2013. "Observing Regional Divergence of Chinese Nanotechnology Centers," Technological Forecasting and Social Change 81: 11–21.

Mowery, David C. 2009. "Plus ca Change: Industrial R&D in the 'Third Industrial Revolution'," Industrial and Corporate Change 18(1): 1–50.

Mozur, Paul and Keith Bradsher. 2017. "China's AI Advances Help Its Tech Industry, and State Security," The New York Times (December 3). Available at https://www.nytimes.com/2017/12/03/business/china-artificial-intelligence.html?_r=0

Mozur, Paul and Mike Isaac. 2016. "Uber to Sell Rival Didi Chuxing and Create New Business in China," New York Times (August 1). Available at https://www.nytimes.com/2016/08/02/business/dealbook/china-uber-didi-chuxing.html

Mullany, Thomas. 2017. The Chinese Typewriter. Cambridge, MA: MIT Press.

Nanopolis, 2015. SIP – Nanopolis website. Available at http://ns.nanopolis.cn/nanopolisEnglish/

National Natural Science Foundation of China (NSFC). 2017. "Yang Wei: Papers Supported by the National Natural Science Foundation of China Accounted for One-Ninth of the Papers Published Globally" (in Chinese). Available at http://www.nsfc.gov.cn/publish/portal0/tab434/info69785.htm

National Natural Science Foundation of China (NSFC). 2018. "NSFC Program Guideline 2018" (in Chinese). Available at http://www.nsfc.gov.cn/nsfc/cen/xmzn/2018xmzn/index.html

National People's Congress (NPC). 2016. "The 13th Five-Year Plan for the National Economic and Social Development of the People's Republic of China" (in Chinese). Available at http://www.npc.gov.cn/wxzl/gongbao/2016-07/08/content_1993756.htm

Nature. 2013. "2013 Tables: Institutions." Nature Index (https://www.natureindex.com/annual-tables/2013/institution/all/all)

Nature. 2017. "2017 Tables: Institutions." Nature Index (https://www.natureindex.com/annual-tables/2017/institution/all/all).

Nederveen Pieterse, Jan. 2015. "China's Contingencies and Globalization," *Third World Quarterly* 36(11): 1985–2001.

Nelson, Richard R. 1959. "The Simple Economics of Basic Scientific Research," *Journal of Political Economy* 67(3): 297-306.

Neubauer, Deane and Jianxin Zhang. 2015. "The Internationalization of Chinese Higher Education," Council for Higher Education Accreditation. Available at https://www.chea.org/userfiles/uploads/Internationalization%20of%20Chinese%20HE-ver2.pdf

Newby, Laura. 2018. *Sino-Japanese Relations: China's Perspective*. New York: Routledge.

Ni Zing and Li Shen. 2016. "Perceptions of Clean Governance: The State of Discrepancy and its Explanation," *Journal of Public Administration (China)*: 3.

Nie, Rongzhen. 1988. *Inside the Red Star: The Memoirs of Marshal Nie Rongzhen*. Beijing: New World Press.

Normile, Dennis and Richard Stone. 2018. "National Science Foundation to Close its Overseas Offices," *Science* (February 26). Available at http://www.sciencemag.org/news/2018/02/national-science-foundation-close-its-overseas-offices

NSF 2014. "Chapter 4. Research and Development: National Trends and International Comparisons," *Science and Engineering Indicators*. Available at https://www.nsf.gov/statistics/seind14/index.cfm/chapter-4/c4s2.htm

NSF and NCSES. 2016. "Doctorate Recipients from US Universities: 2016." Available at https://www.nsf.gov/statistics/2018/nsf18304/data.cfm

NSTC. 2000. "National Nanotechnology Initiative: Leading to the Next Industrial Revolution." Available at: https://clintonwhitehouse3.archives.gov/WH/EOP/OSTP/NSTC/html/iwgn/iwgn.fy01budsuppl/nni.pdf

OECD. 2013. *The People's Republic of China: Avoiding the Middle Income Trap – Policies for Sustained and Inclusive Growth*. Paris: OECD Publishing. Available at http://www.oecd.org/publications/the-people-s-republic-of-china-avoiding-the-middle-income-trap-policies-for-sustained-and-inclusive-growth-9789264207974-en.htm

OECD. 2014. "China Headed to Overtake EU, US in Science & Technology Spending, OECD Says" (December 11). Available at http://www.oecd.org/newsroom/china-headed-to-overtake-eu-us-in-science-technology-spending.htm

OECD. 2017a. "Financial Indicators – Stocks: Private Sector Debt," OECD. Stat. Available at http://stats.oecd.org/index.aspx?queryid=34814

OECD. 2017b. *The Size and Sectoral Distribution of State-Owned Enterprises*. Paris: OECD Publishing.

OECD MSTI. 2017. "Main Science and Technology Indicators." Available at http://www.oecd.org/sti/msti.htm

OECD/EUIPO. 2016. *Trade in Counterfeit and Pirated Goods: Mapping the Economic Impact*. Paris: OECD Publishing.

Ong, Ryan. 2009. "Tackling Intellectual Property Infringement in China," *China Business Review* 36(2). Available at https://www.chinabusinessreview.com/tackling-intellectual-property-infringement-in-china/

Orleans, Leo. 1988. *Chinese Students in America: Policies, Issues, and Numbers*. Washington, DC: National Academy Press.

Orlik, Tom. 2013. "China Labors On as Wages Leap Higher," *Wall Street Journal on-line* (March 15). Available at http://online.wsj.com/article/SB100014241 27887324077704578361351172124228.html

Ouyang, Yadan. 2017. "The Respiratory Landscape in China: SA Focus on Air Pollution," *The Lancet* 5(1): 16–17.

Parker, Emily. 2017. "Can WeChat Thrive in the United States?" *MIT Technology Review* (August 11). Available at https://www.technologyreview.com/s/608578/can-wechat-thrive-in-the-united-states/

Parker, Rachel and Richard Appelbaum. 2012. "The Chinese Century? Some Implications of China's Move to High-Tech Innovation for US Policy," in *The Social Life of Nanotechnology*, edited by Barbara Herr Harthorn and John Mohr, 134–65. New York: Routledge.

Peck, Jamie and Jun Zhang. 2013. "A Variety of Capitalism ... With Chinese Characteristics?" *Journal of Economic Geography* 13(3): 357–96.

People.cn. 2015. "International Cooperation Papers Account for around 25% of Total" (in Chinese). Available at http://scitech.people.com.cn/n/2015/1021/c1007-27723602.html

Pereira, Alexius A. 2003. *State Collaboration and Development Strategies in China: The Case of the China–Singapore Suzhou Industrial Park (1992–2002)*. New York: RoutledgeCurzon.

Perez, Bien. 2017. "China Closes Gap with US in Hi-Tech Breakthroughs, KPMG Finds," *South China Morning Post* (March 6). Available at http://www.scmp.com/tech/innovation/article/2076348/china-closes-gap-us-hi-tech-breakthroughs-kpmg-finds

Pettis, Michael. 2013. *Avoiding the Fall: China's Economic Restructuring*. Washington, DC: Carnegie Endowment for International Peace.

Polanyi, Karl. 1957. *The Great Transformation: The Political and Economic Origins of our Time*. Boston, MA: Beacon Press.

Poo, Mu-ming and Ling Wang. 2014. "On CAS Pioneer Initiative – An Interview with CAS President Chunli Bai," *National Science Review* 1: 618–22.

PRNewswire. 2015. "Haidian Park: The Birthplace of China's Most Innovative and Entrepreneurial Technology Companies." Available at http://www.prnewswire.

com/news-releases/haidian-park-the-birthplace-of-chinas-most-innovative-and-entrepreneurial-technology-companies-300024231.html

Puffer, Sheila M., Daniel J. McCarthy, and Max Boisot. 2010. "Entrepreneurship in Russia and China: The Impact of Formal Institutional Voids," *Entrepreneurship Theory and Practice* 34: 441–67.

Qiu, Jane. 2016. "Nanotechnology Development in China: Challenges and Opportunities," *National Science Review* 3: 148–52.

Qualcomm. 2017. "LTE IoT is Starting to Connect the Massive IoT Today, Thanks to eMTC and NB-IoT." Available at https://www.qualcomm.com/news/onq/2017/06/15/lte-iot-starting-connect-massive-iot-today-thanks-emtc-and-nb-iot

Quandl. 2016. "China Total Investment, % of GDP," IMF Cross Country Macroeconomic Database (October 6). Available at https://www.quandl.com/data/ODA/CHN_NID_NGDP-China-Total-Investment-of-GDP

Ramzy, Austin, 2017. "Chinese Court Sentences Activist Who Documented Protests to 4 Years in Prison," *The New York Times* (August 4). Available at https://www.nytimes.com/2017/08/04/world/asia/china-blogger-lu-yuyu-prison-sentence-protests-picking-quarrels.html?mcubz=0&_r=0

Rao Yi, Daqing Zhang, and Runhong Li. 2015. *Tu Youyou and Artemisinin* (in Chinese). Beijing: China Science and Technology Publisher.

Resnik, David and Weiqin Zeng. 2010. "Research Integrity in China: Problems and Prospects." *Developing World Bioethics* 10(3): 164–71.

Reuters. 2016. "China's Yuan Joins Elite Club of IMF Reserve Currencies," Reuters (September 30). Available at https://www.reuters.com/article/us-china-currency-imf/chinas-yuan-joins-elite-club-of-imf-reserve-currencies-id USKCN1212WC

Roco, Mihail C. 2007. 'National Nanotechnology Initiative: Past, Present, Future', in *Handbook of Nanoscience, Engineering, and Technology*, 2nd edn, edited by W.A. Goddard, D. W. Brenner, S. E. Lyshevsky and G. J. Iafrate, section 1, chapter 3. Boca Raton, FL: CRC Press.

Roco, Mihail C., S. Williams, and P. Alivisatos. 1999. "Nanotechnology Research Directions: IWGN Workshop Report – Vision for Nanotechnology Research and Development in the Next Decade," International Technology Research Institute, World Technology (WTEC) Division, Loyola University, Baltimore, MD.

Rodríguez-Pose, Andrés and Daniel Hardy. 2014. *Technology and Industrial Parks in Emerging Countries*. New York: Springer.

Rotenberg, Ziv. 2015. "China's New Intellectual Property Courts." Lexology (March 5). Available at https://www.lexology.com/library/detail.aspx?g=9c524ea5-4ab3-4eec-87f7-08c5307fc332

Rothman, Andy. 2015. "Sinology: The Coming Chinese Crackup?" Matthews Asia (March 31). Available at https://us.matthewsasia.com/perspectives-on-asia/sinology/article-914/default.fs

Samuelson, Robert J. 2018. "China's Breathtaking Transformation into a Scientific Superpower," *Washington Post* (January 21). Available at https://www.washingtonpost.com/opinions/chinas-breathtaking-transformation-into-a-scientific-superpower/2018/01/21/03f883e6-fd44-11e7-8f66-2df0b94bb98a_story.html?utm_term=.bcadc1c526ed

Saxenian, Annalee. 1994. *Regional Advantage*. Cambridge, MA: Harvard University Press.

SCMP. 2017. "China's Forex Reserves Hit 11-Month High of US$3.1 Trillion," *South China Morning Post* (October 9). Available at http://www.scmp.com/news/china/economy/article/2114579/chinas-forex-reserves-hit-11-month-high-us31-trillion

Segal, Adam. 2003. *Digital Dragon: High-Technology Enterprises in China*. Ithaca, NY: Cornell University Press.

Segal, Adam. 2011. *Advantage: How American Innovation Can Overcome the Asian Challenge*. New York: W.W. Norton & Company.

Sergi, Brian, Rachel Parker, and Brian Zuckerman. 2014. "Support for International Collaboration in Research: The Role of the Overseas Offices of Basic Science Funders," *Review of Policy Research* 31(5): 430–53.

Shambaugh, David. 2015. "The Coming Chinese Crackup," *The Wall Street Journal* (March 6). Available at https://www.wsj.com/articles/the-coming-chinese-crack-up-1425659198

Shambaugh, David and Eberhard Sandschneider. 2007. *China–Europe Relations: Perceptions, Policies and Prospects*. New York: Routledge.

Shen, Lan. 2014. "Transformation a Continuous Process at Suzhou Industrial Park," *Business Wire* (September 12). Available at http://www.businesswire.com/news/home/20140912005095/en/Transformation-Continuous-%20Process-Suzhou-Industrial-Park

Shen, Lan. 2015. "Suzhou Industrial Park: The Pilot for Change," *Business Wire* (October 13). Available at http://www.businesswire.com/news/home/20151013005734/en/Suzhou-Industrial-Park-Pilot-Change

Shen, Zhihua and Yafeng Xia. 2012. "Between Aid and Restriction: Changing Soviet Policies Toward China's Nuclear Weapons Program, 1954–1960. Nuclear Proliferation International Project, Wilson Center, Working Paper 2:1–89.

Shen, Zhihua and Yafeng Xia. 2015. *Mao and the Sino–Soviet Partnership, 1945–1959*. Lanham, MD: Lexington Books.

Shenzhen Daily. 2017. "Forex Reserves at 9-Month High" (August 8). Available at http://www.szdaily.com/content/2017-08/08/content_16940304.htm

Shi, Han, Jinping Tian, and Luqun Chen. 2012. "China's Quest for Eco-Industrial Parks, Part I: History and Distinctiveness," *Journal of Industrial Ecology* 16:1: 8–10.

Shi, Yigong and Yi Rao. 2010. "China's Research Culture," *Science* 329(5996): 1128.

Shira, Dezan. 2011. "MOFCOM Ranks China's Top Development Zones," China Briefing: Business Intelligence from Dezan Shira and Associates (September 29). Available at http://www.china-briefing.com/news/2011/09/29/mofcom-ranks-chinas-top-development-zones.html

Simon, Denis Fred and Cong Cao. 2009a. "Creating an Innovative Talent Pool," *China Business Review* (November 2). Available at http://www.chinabusiness-review.com/creating-an-innovative-talent-pool/

Simon, Denis Fred and Cong Cao. 2009b. *China's Emerging Technological Edge: Assessing the Role of High-End Talent.* Cambridge: Cambridge University Press.

SIP. 2009. "Learn from the World" website, Suzhou Industrial Park. Quoted from *Xinhua Daily*, May 24. Available at http://www.sipac.gov.cn/english/zhuanti/jg60n/gjlnbtsj/

SIP. 2013. "Suzhou Industrial Park Innovates Nanotech Development Pattern: An Overview of No. 1 Industry in SIP" (September 29). Available at http://www.sipac.gov.cn/english/categoryreport/IndustriesAndEnterprises/201309/t20130929_235576.htm

SIP. 2014. "Two Decades of China–Singapore Cooperation Creating a Beautiful New Paradise in SIP," SIP Special Report on 20th Anniversary of China–Singapore Suzhou Industrial Park." Available at http://www.sipac.gov.cn/english/zhuanti/20140429yq20zn/zxhz/

SIP. 2016. "Learning from the World." Available at http://www.sipac.gov.cn/english/zhuanti/fnotpoc/fnotpoc_lftw/

SIP BioBay. 2008. "Live, Breathe, Innovate and Succeed," Slide Deck on Suzhou Industrial Park Bio Bay. Available at https://www.slideshare.net/zhangdc/suzhou-industrial-park-bio-bay

SIP Nanopolis. 2018. Nanopolis Suzhou Co., Ltd website. Available at http://www.nanopolis.cn/en/Index.aspx

SIPAC. 2015. *Suzhou Industrial Park Scientific Masterplan*, SIP Administrative Committee. Available at http://www.sipac.gov.cn/english/InvestmentGuide/ScientificMasterPlan/201107/t20110704_102990.htm

SIPAC. 2018. "Investment Guide," China-Singapore Suzhou Industrial Park website. Available at www.sipac.gov.cn/english/InvestmentGuide/SinoSingaporeCooperation/201107/t20110704_102985.htm

SIPO. 2017. "Intellectual Property Protection Highlighted on the Two Sessions." Available at http://english.sipo.gov.cn/news/official/201703/t20170315_1308796.html

SNC. 2011. "Nanotechnology Capabilities Report of Suzhou, China," Suzhou Nanotech Co. Ltd & NanoGlobe Pte. Ltd. Available at http://www.nanoglobe.biz/News/SuzhouReport_English_2nd_e-version.pdf

Soper, Taylor. 2017. "Chinese Tech Giant Tencent is Poised to Be a Leader in AI, Says Head of New Seattle Research Lab," *GeekWire* (December 14). Available at https://www.geekwire.com/2017/chinese-tech-giant-tencent-poised-leader-ai-says-head-new-seattle-research-lab/

Sparks, Daniel. 2017. "How Many Users Does Twitter Have?" *The Motley Fool* (April 27). Available at https://www.fool.com/investing/2017/04/27/how-many-users-does-twitter-have.aspx

Springut, Micah, Stephen Schlaikjer, and David Chen. 2011. "China's Program for Science and Technology Modernization: Implications for American Competitiveness," Washington, DC: The US–China Economic and Security Review Commission.

State Council. 2017a. "China to Invest Big in 'Made in China 2025' Strategy." Available at http://english.gov.cn/state_council/ministries/2017/10/12/content_281475904600274.htm

State Council. 2017b. "World Factory Embraces Intelligent Manufacturing." Available at http://english.gov.cn/news/video/2017/09/12/content_281475852344964.htm

State Council of the PRC. 2015. "Made in China 2025" (in Chinese). Available at http://www.gov.cn/zhengce/content/2015-05/19/content_9784.htm

Steinfeld, Edward S. 2010. *Playing Our Game: Why China's Rise Doesn't Threaten the West.* Oxford: Oxford University Press.

Stephan, Paula E. 2011. *How Economics Shapes Science.* Cambridge, MA: Harvard University Press.

Stone, Brad and Lulu Yilun Chen. 2016. "Uber Slayer: How China's Didi Beat the Ride-Hailing Superpower," *Bloomberg Businessweek* (October 7). Available at https://www.bloomberg.com/features/2016-didi-cheng-wei/

Stone, Brad and Lulu Yilun Chen. 2017. "Tencent Dominates in China. Next Challenge Is Rest of the World," *Bloomberg Businessweek* (June 28). Available at https://www.bloomberg.com/news/features/2017-06-28/tencent-rules-china-the-problem-is-the-rest-of-the-world

Storper, Michael and Anthony J. Venables. 2004. "Buzz: Face-to-Face Contact and the Urban Economy," *Journal of Economic Geography* 4(4): 351–70.

Sun, Yifei, Max von Zedtwitz, and Denis Fred Simon (eds.) 2013. *Global R&D in China*. New York: Routledge.

Sun, Yiting. 2017. "Why 500 Million People in China Are Talking to This AI," *MIT Technology Review* (September 14). Available at https://www.technologyreview.com/s/608841/why-500-million-people-in-china-are-talking-to-this-ai/

Sun, Yutao and Cong Cao. 2014. "Demystifying Central Government R&D Spending in China," *Science* 345(6200): 1006–8.

Sun, Yutao and Seamus Grimes. 2017. *China and Global Value Chains: Globalization and the Information and Communications Technology Sector*. London: Routledge.

Suttmeier, Richard P. 2007. "Engineers Rule, OK?" *New Scientist*, November 10, pp. 71–73.

Suttmeier, Richard P. 2014a. Co-inventing the Future? Science Diplomacy and the Evolution of Sino–US Relations in Science and Technology. Discussion paper presented at conference entitled "The Evolving Role of Science and Technology in China's International Relations," April 3–4, Arizona State University, Tempe, AZ.

Suttmeier, Richard P. 2014b. "Trends in US–China S&T Cooperation: Collaborative Knowledge Production for the Twenty-First Century." Research report prepared for the US–China Economic and Security Review Commission, Washington, DC (September 11).

Suttmeier, Richard P. and Cong Cao. 2006. China–US S&T Cooperation: Past Achievements and Future Challenges. US–China Forum on Science and Technology Policy. Available at http://china-us.uoregon.edu/pdf/CHIN-AUSSTCOOPERATIONForum.pdf

Suttmeier, Richard P. and Denis Fred Simon. 2014. "Conflict and Cooperation in the Development of US–China Relations in Science and Technology: Empirical Observations and Theoretical Implications," in *The Global Politics of Science and Technology*, Volume 2, edited by M. Mayer, M. Carpes, and R. Knoblich. Berlin: Springer-Verlag.

Suttmeier, Richard P., Cong Cao, and Denis Fred Simon. 2006a. "'Knowledge Innovation' and the Chinese Academy of Sciences," *Science* 312(5770): 58–9.

Suttmeier, Richard P., Cong Cao, and Denis Fred Simon. 2006b. "China's Innovation Challenge and the Remaking of the Chinese Academy of Sciences," *Innovations: Technology, Governance, Globalization* 1(3): 78–97.

Suzhou Industrial Park. 2011. Nanotechnology Capabilities Report of Suzhou, China. Suzhou Nanotech Co. Ltd. Available at www.nano-globe.biz/News/SuzhouReport_English_2nd_e-version.pdf

Swanstrom, Niklas and Ryosei Kokubun. 2012. *Sino–Japanese Relations: Rivals or Partners in Regional Cooperation*. Singapore: World Scientific.

Tencent. 2017. "About Tencent." Available at https://www.tencent.com/en-us/abouttencent.html

Tencer, Daniel. 2017. "Half of China's Millionaires Want to Emigrate, and Canada's Their #2 Choice," *Huffington Post* (July 17). Available at http://www.huffingtonpost.ca/2017/07/17/half-of-china-s-millionaires-want-to-emigrate-and-canada-s-thei_a_23034094/

Tilley, Aaron. 2016. "Xiaomi Takes Its First Real Step into the US Market," *Forbes* (October 13). Available at https://www.forbes.com/sites/aarontilley/2016/10/13/xiaomi-takes-its-first-real-step-into-the-us-market/#6eb622b562ac

Torch Program. 2011. *National High-Tech Industrial Zones in China.* Ministry of Science and Technology, Torch High Technology Industry Development Center. People's Republic of China.

Trading Economics. 2016. "China Foreign Exchange Reserves, 1980–2016." http://www.tradingeconomics.com/china/foreign-exchange-reserves

Trading Economics. 2017a. "China average yearly wages in manufacturing." Available at https://tradingeconomics.com/china/wages-in-manufacturing

Trading Economics. 2017b. "China Foreign Exchange Reserves, 1980–2017." Available at https://tradingeconomics.com/china/foreign-exchange-reserves

Unger, Jonathan, and Anita Chan. 1995. "China, Corporatism, and the East Asian Model," *Australian Journal of Chinese Affairs* 33: 25–93.

United States Conference of Mayors. 2017. "Cities of the 21st Century: 2016 Smart Cities Survey." Available at https://www.usmayors.org/wp-content/uploads/2017/02/2016SmartCitiesSurvey.pdf

US Chamber of Commerce. 2017. "Made in China 2025: Global Ambitions Built on Local Protections." Available at https://www.uschamber.com/sites/default/files/final_made_in_china_2025_report_full.pdf

US Department of Treasury. 2017. "Major Foreign Holders of Treasury Securities." Available at http://ticdata.treasury.gov/Publish/mfh.txt

US FRB. 2017a. "Major Foreign Holdings of Treasury Securities," US Federal Reserve Board, Department of the Treasury (August 15). Available at http://ticdata.treasury.gov/Publish/mfh.txt

US FRB. 2017b. "Major Foreign Holdings of Treasury Securities," US Federal Reserve Board, Department of the Treasury, 2000–2016 (August 15). Available at http://ticdata.treasury.gov/Publish/mfhhis01.txt

US NAS. 2016. *Triennial Review of the National Nanotechnology Initiative.* National Academy of Sciences: The National Academies Press.

US–China Business Council. 2013. "China's Strategic Emerging Industries: Policy, Implementation, Challenges, & Recommendations." Available at https://www.uschina.org/sites/default/files/sei-report.pdf

US–China Clean Energy Research Center. 2013. "Annual Report 2012–2013: Accomplishment from the Second Year of the US–China Clean Energy Research Center." Available at http://www.us-china-cerc.org/pdfs/US-China_CERC_Annual_Report_2012-2013.pdf

USTR. 2017. "Special 301 Report on Intellectual Property Rights." Office of the United States Trade Representative. Available at https://ustr.gov/sites/default/files/301/2017%20Special%20301%20Report%20FINAL.PDF

Van Noorden, Richard. 2014. "Publishers Withdraw More Than 120 Gibberish Papers," *Nature* (February 24). DOI: 10.1038/nature.2014.14763

Van Winden, Willem, Erik Braun, Alexander Otgaar, and Jan-Jelle Witte. 2014. *Urban Innovation Systems: What Makes Them Tick?* London: Routledge.

Veldhoen, Steven, Bill Peng, and Anna Mansson. 2014. *2014 China Innovation Survey: China's Innovation is Going Global.* PWC (September 23). Available at http://www.strategyand.pwc.com/reports/2014-china-innovation-survey

Vodafone. 2017. "Narrowband-IoT: Connecting the Internet of Hidden Things." Available at: http://www.vodafone.com/business/iot/nb-IoT.

Wagner, Caroline S., Lutz Bornmann, and Loet Leydesdorff. 2015. "Recent Developments in China–US Cooperation in Science," *Minerva* 53(3): 199–214.

Walsh, Kathleen. 2003. *Foreign High-Tech R&D in China.* Washington, DC: Henry L. Stimson Center.

Wan, Gang. 2018. "Remarks at Two Session Press Conference Regarding Speeding Up Construction of an Innovation Country" (March 10). Available at http://www.stdaily.com/app/yaowen/2018-03/10/content_646178.shtml

Wang, Hongyi. 2016. "Smart City Project Transforms Residents' Lives," *China Daily* (October 31). Available at http://usa.chinadaily.com.cn/epaper/2016-10/31/content_27227164.htm

Wang, Hui. 2017. "Highest Damages Ever Rewarded by Beijing IP Court and Stronger Patent Protection in China," Lexology (July 21). Available at: https://www.lexology.com/library/detail.aspx?g=a31bfaf8-c489-41f4-b109-df4c2dd75f97

Wang, Huiyao, David Zweig, and Xiaohua Lin. 2011. "Returnee Entrepreneurs: Impact on China's Globalization Process." *Journal of Contemporary China* 20: 413–31.

Wang, Haiyan, Richard Appelbaum, Francesca de Giuli, and Nelson Lichtenstein. 2009. "China's New Contract Labor Law: Is China Moving Towards Increased Power for Workers?" *Third World Quarterly* 30(3): 485–501.

Wang, Qi. 2017. "Programmed to Fulfill Global Ambitions," *Nature* 545, May 25, p. S53.

Wang, Yue. 2017a. "Will the Future of Artificial Intelligence Look Chinese?" *Forbes* (November 6). Available at https://www.forbes.com/sites/ywang/2017/11/06/will-the-future-of-artificial-intelligence-look-chinese/#74973b667fdc

Wang, Yue. 2017b. "Xiaomi Is Once Again Among the World's Biggest Smartphone Brands – But Could It Go Further?" *Forbes* (August 8). Available at https://www.forbes.com/sites/ywang/2017/08/08/xiaomi-is-once-again-worlds-biggest-smartphone-brand-but-could-it-go-further/#673829711691

WEF. 2016. *The Global Competitiveness Report 2016–2017*. WEF, Columbia University. Available at http://www3.weforum.org/docs/GCR2016-2017/05FullReport/TheGlobalCompetitivenessReport2016-2017_FINAL.pdf

Wei, Y.H. Dennis, Yuqi Lu, and Wen Chen. 2009. "Globalizing Regional Development in Sunan, China: Does Suzhou Industrial Park Fit a Neo-Marshallian District Model?" *Regional Studies* 43(3): 409–27.

Weinberger, Matt. 2016. "Uber to Merge with Chinese Rival Didi in $35 Billion Deal," *Business Insider* (August 1). Available at http://www.businessinsider.com/uber-china-merges-with-didi-chuxing-2016-7

Weiss, Linda. 2014. *America Inc.? Innovation and Enterprise in the National Security State*. Ithaca, NY: Cornell University Press.

Wen, Cong. 2017. "Dongguan: Science and Technology Week Promotes Docking of Environmental Projects of Chinese and Japanese Enterprises," Sina.com (December 9). Available at http://k.sina.com.cn/article_1497087080_593b bc68042003v99.html

Wen, Gong, Xu Zhefeng, and Wu Qiuyu. 2016. "Accessing the General Trend of the First-Quarter: Authorities on Current Chinese Economy" (in Chinese), *People's Daily*, May 9, p. 1.

Wertime, David. 2014. "It's Official: China is Becoming a New Innovation Powerhouse," *Foreign Policy* (February 6). Available at http://foreign-policy.com/2014/02/07/its-official-china-is-becoming-a-new-innovation-powerhouse/

Wike, Richard and Bridget Parker. 2015. "Corruption, Pollution, Inequality Are Top Concerns in China," Pew Research Center (September 24). Available at http://www.pewglobal.org/2015/09/24/corruption-pollution-inequality-are-top-concerns-in-china/

Wilson, Jeanne. 2004. *Strategic Partners: Russian–Chinese Relations in the Post-Soviet Era*. New York: Routledge.

WIPO. 2016a. *Global Innovation Index 2016*. Geneva: World Intellectual Property Organization. Available at http://www.wipo.int/edocs/pubdocs/en/wipo_pub_gii_2016.pdf

WIPO. 2016b. *World Intellectual Property Indicators 2016*. Geneva: World Intellectual Property Organization. Available at http://www.wipo.int/edocs/pubdocs/en/wipo_pub_941_2016.pdf

Wong, Joseph. 2011. *Betting on Biotech: Innovation and the Limits of Asia's Developmental State*. Ithaca, NY: Cornell University Press.

Woodruff, Betsy and Julia Arciga. 2018. "FBI Director's Shock Claim: Chinese Students Are a Potential Threat," *Daily Beast*, February 13. Available at https://www.thedailybeast.com/fbi-directors-shock-claim-chinese-students-are-a-potential-threat

World Bank. 2013. China 2030: Building a Modern, Harmonious and Creative Society. Washington, DC: World Bank. Available at https://openknowledge.worldbank.org/bitstream/handle/10986/12925/9780821395455.pdf?sequence=5

World Bank. 2017. "Spotlight 6: The Middle Income Trap," in *World Development Report 2017: Governance and the Law*, 159–62. Washington, DC: World Bank. Available at http://www.worldbank.org/en/publication/wdr2017

Wu, Qiang. 2016. "What Do Lu Yuyu's Statistics of Protest Tell Us About the Chinese Society Today?" *China Change* (July 6). Available at https://chinachange.org/2016/07/06/the-man-who-keeps-tally-of-protests-in-china/

Xi Jinping. 2012. "Exploring the Vast Universe Hand in Hand, Working Together Toward a Better Future for Humankind," *Inquiries of Heaven*, IAU 28th General Assembly, Beijing, China (August 22). Available at http://info.bao.ac.cn/download/astronomy/IAU2012/newspaper/IHissue03.pdf

Xie, Yu, Chunni Zhang, and Qing Lai. 2014. "China's Rise as a Major Contributor to Science and Technology," *Proceedings of the National Academy of Sciences* 111(26): 9437–42.

Xinhua. 2013. "Politburo Holds Its Group Study outside Zhongnanhai for the First Time" (October 1, in Chinese). Available at http://news.xinhuanet.com/politics/2013-10/01/c_117582235.htm

Xinhua. 2015. "China–Singapore Suzhou Industrial Park Targets New Reforms" (October 28). Available at http://www.chinadaily.com.cn/business/2015-10/28/content_22302520.htm

Xinhua. 2016. "The Outline of the Thirteenth Five-Year Plan for the National Economy and Social Development of the People's Republic of China" (March 17, in Chinese). Available at http://news.xinhuanet.com/politics/2016lh/2016-03/17/c_1118366322.htm

Xinhua. 2017a. "Li Keqiang: Mass Entrepreneurship and Innovation Is Flourishing," *Xinhuanet* (September 12). Available at http://www.xinhuanet.com/english/2017-09/12/c_136603727.htm

Xinhua. 2017b. "China to Further Promote Innovation, Entrepreneurship," *Xinhuanet* (July 12). Available at http://www.xinhuanet.com/english/2017-07/12/c_136438984.htm

Xu, Ang. 2008. "China Looks Abroad: Changing Direction in International Science." *Minerva* 46(1): 37–51.

Xu, Chenggang. 2015. "China's Political-Economic Institutions and Development," *Cato Journal* 35(3): 525–48.

Yahuda, Michael. 2013. *Sino–Japanese Relations After the Cold War*. New York: Routledge.

Yam, Kimberly. 2018. "FBI Director Defends Remarks That Chinese People in US Pose Threats," *Huffington Post*, May 23. Available at https://www. huffingtonpost.com/entry/fbi-christopher-wray-chinese-immigrants_us_ 5ab3d47fe4b008c9e5f51975

Yang, Shengping. 2011. "Patent Enforcement in China," *Landslide* 4(2). Available at http://www.americanbar.org/content/dam/aba/publications/landslide/ landslide_november_2011/yang_landslide_novedec_2011.authcheckdam.pdf

Yiu, Enoch. 2016. "China's Booming Start-Up Culture Lures Venture Capitalists in Search of the Next US$1b Unicorns," *South China Morning Post* (May 18). Available at http://www.scmp.com/business/article/1946763/ chinas-booming-start-culture-lures-venture-capitalists-search-next-us1b

Zacharakis, Andrew L., Jeffrey S. McMullen, and Dean A. Shepherd. 2007. "Venture Capitalists' Decision Policies across Three Countries: An Institutional Theory Perspective," *Journal of International Business Studies* 38(5): 691– 708.

Zacks Equity Research. 2017. "Are LTE Cat-M and NB-IoT the Next Big Things in Wireless?" Available at http://www.nasdaq.com/article/ are-lte-cat-m-and-nb-iot-the-next-big-things-in-wireless-cm853820

Zhan, Ying. 2014. "Problems of Enforcement of Patent Law in China and Its Ongoing Fourth Amendment," *Journal of Intellectual Property Rights* 19: 266–71.

Zhao, Min and Thomas Farole. 2011. "Partnership Arrangements in the China– Singapore (Suzhou) Industrial Park: Lessons for Joint Economic Zone Development," in *Special Economic Zones: Progress, Emerging Challenges, and Future Directions*, edited by Thomas Farole and Gokhan Akinci. Washington, DC: The World Bank.

Zhao, Zhang. 2012. "Pioneering Role in Intellectual Property Protection," *China Daily* (January 18). Available at http://ipr.chinadaily.com.cn/2012-01/18/ content_14470786.htm

Zhou, Yu. 2008. "China's High Tech Industry and the World's Economy: Zhongguancun Park," *The Asia-Pacific Journal* 6(2): 1–12.

Zhou, Yu, William Lazonick, and Yifei Sun (eds.) 2016. *China as an Innovation Nation*. Oxford: Oxford University Press.

Zhuo Wenting. 2017. "Shanghai Unveils Steps to Attract Foreign R&D Centers," *China Daily* (October 17). Available at http://www.chinadaily.com.cn/business/2017-10/17/content_33352161.htm

Zimmerman, James. 2015. "The 'New Normal' for Doing Business in China," *The Wall Street Journal* (February 10). Available at https://www.wsj.com/articles/james-zimmerman-the-new-normal-for-doing-business-in-china-1423587712

Zweig, David. 2006. "Learning to Compete: China's Efforts to Encourage a Reverse Brain Drain." *International Labour Review* 145 (1–2): 65–90.

Zweig, David and Huiyao Wang. 2013. "Can China Bring Back the Best? The Communist Party Organizes China's Search for Talent." *The China Quarterly* 215: 590–615.

Zukus, Jason. 2017. "Globalization with Chinese Characteristics: A New International Standard," *The Diplomat* (May 9). Available at http://thediplomat.com/2017/05/globalization-with-chinese-characteristics-a-new-international-standard/

Index